PILING

Model procedures and specifications

INSTITUTION OF CIVIL ENGINEERS, LONDON, 1978

Published by the Institution of Civil Engineers, Telford House, P.O. Box 101, 26–34 Old Street, London EC1P 1JH

ISBN 0 7277 0036 7

Produced and distributed by Thomas Telford Ltd, Telford House, P.O. Box 101, 26–34 Old Street, London EC1P 1JH

Typeset at The Pitman Press, Bath
Printed and bound by William Clowes & Sons, Limited, London, Beccles and Colchester

PREFACE

M. W. LEONARD, Chairman, ICE Piling (Group) Committee 1968–75

The Council of the Institution of Civil Engineers in 1966 set up a Piling Committee—later to become the Piling Group Committee—to undertake a wide range of activities in the specialist field of piling foundations. The committee comprised representatives from consultants, contractors and government and public bodies, and prepared an on-going comprehensive programme for piling and related foundations covering research and development, model specifications, data retrieval and a series of informal discussions and conferences for piling practitioners.

Recently the research and development proposals of the committee were taken up by the Construction Industry Research and Information Association who are carrying out work in conjunction with the Property Services Agency and the Building Research Establishment under contract with the Department of the Environment. The informal discussions continued to provide a general forum for the exchange of data and experience on everyday problems for all those concerned with piling.

By far the largest and most difficult task was undertaken by the Piling Specifications Committee, set up in 1969 to draft model specifications and special conditions for piling work, which were felt by those engaged in such work to be very necessary to facilitate satisfactory performance of this important aspect of civil engineering practice. This Committee, under the Chairmanship of David Palmer, has persevered over several years to meet the diverse needs of the piling industry. Albeit a compromise in parts to overcome hindrances to its completion, the present resulting creditable document will, I feel, be welcomed by all.

So much in piling depends on an exercise of judgement by the experienced engineer; performance depends on sound and alert supervision of the actual works. All this can impose a near intolerable burden on those responsible if they have in addition to meet a specification and contract terms which are confusing and contradictory.

Having spent much of my career in piling, I feel confident that this publication will be invaluable as an authoritative guide to the written element of the obligations to be entered into by Employer and Contractor. For the Engineer to the Contract, it will provide a practical aid to enable him to discharge his part in line with the needs of the piling works.

PILING SPECIFICATIONS COMMITTEE

F. R. Bullen, BSc, FICE, FIStructE (Chairman to 1969*); D. J. Palmer, MA, FICE (Chairman from 1969); D. Dennington, BSc(Eng), ACGI, FICE, FIStructE; C. K. Haswell, BSc(Eng), FICE, FIStructE; M. W. Leonard, BSc(Eng), FICE, MIMechE; F. A. Sharman, BSc(Eng), ACGI, FICE, MInstHE, FRGS; T. Whitaker, DSc, MICE

DRAFTING PANELS

Paragraphs 37–59 and clauses 1.01–7.12

S. Packshaw, BSc, FICE (Chairman to 1969*); D. J. Palmer, MA, FICE (Chairman from 1969); F. H. Armitage, BEng, MICE; F. G. Coffin, FIStructE (from 1970); P. A. Cox, BSc(Eng), ACGI, FICE; N. G. Eggleton, BSc(Eng), FICE; W. G. K. Fleming, PhD, MICE; L. R. Greenaway, OBE, BSc(Eng), FICE (to 1972); G. P. Mallett, BSc(Eng), ACGI, MICE (from 1972); E. Newton, BSc, MICE (1971–73); F. A. Page, BSc(Eng), FICE, MIStructE (1969–71); J. H. Sherry, BSc(Eng), MICE

Paragraphs 60–98 and clauses 8.01–8.186

F. R. Bullen, BSc, FICE, FIStructE (Chairman to 1969*); T. Whitaker, DSc, MICE (Chairman from 1969); D. D. J. Clarke, MBE, BSc(Eng), MICE, FIStructE (to 1970); G. M. Cornfield, MSc(Eng), FICE, MIStructE (to 1974*); J. Dufton, MIStructE, FRICS (from 1970); W. G. K. Fleming, PhD, MICE; J. F. Levy, BSc(Eng), DIC, FIStructE; V. J. F. Weeks, MICE

Paragraphs 1–36 and 99–156, draft preamble and suggested measured items for a bill of quantities

W. Calderwood, BSc, MICE (Chairman to 1969); D. Dennington, BSc(Eng), ACGI, FICE, FIStructE (Chairman from 1969); N. W. Chisholm (from 1971); T. W. Dawkins, FIStructE (to 1969); J. A. Derrington, BSc(Eng), DIC, FICE, FIStructE; E. C. F. Iggulden, FRICS; D. G. Jobling, BSc(Eng), FICE; M. Nachshen, BSc(Eng), FICE, FIStructE (from 1971); D. A. Rolt; A. S. West, FICE (from 1971)

Paragraphs SI 1–SI 59 and clauses SI 1.01–SI 4.02

F. A. Sharman, BSc(Eng), ACGI, FICE, MInstHE, FRGS (Chairman); K. O. Pook, BSc(Eng), MICE; M. J. Tomlinson, FICE, FIStructE

* Deceased.

CONTENTS

Foreword vii

Model procedures for piling 1
    Responsibility for piling 2
    Appointment of contractors and sub-contractors for piling 4
    Use of the model specification for piling 17
    Pile testing 22
    Measurement of piling work 30
    Draft preamble to a bill of quantities 39
    Suggested measured items for a bill of quantities 42

Model specification for piling 53
    Section 1. General requirements for piling work 54
    Section 2. General requirements for concrete piles 62
    Section 3. Precast normal reinforced and prestressed piles 70
    Section 4. Bored cast in place piles 78
    Section 5. Driven cast in place piles 85
    Section 6. Steel piles 92
    Section 7. Timber piles 102
    Section 8. Pile testing 108

Model procedures for site investigation for piling 123
    General 124
    Terminology 124
    Particular clauses 125
    Boring and sampling 127
    In situ testing 131
    Laboratory testing 136

Model specification for site investigation for piling 137
    Section SI 1. Particular clauses 138
    Section SI 2. Boring and sampling 140
    Section SI 3. In situ testing 147
    Section SI 4. Laboratory testing 160

FOREWORD

Since its formation in 1966 the discussions held by the Piling Group of the Institution of Civil Engineers with all sides of the piling industry have aroused great interest and have shown the need for a widely accepted set of specifications for piling work together with particular conditions of contract where it is necessary to supplement the standard forms in general use. Not the least reason for this is that piling work may be unseen once it is constructed and yet it is of the greatest importance to the structure which it supports.

In 1968 a Piling Specifications Committee was set up under terms of reference approved by the Research and Development Committee of the Institution of Civil Engineers to prepare model clauses for use with conditions of contract, model specifications and methods of measurement applicable to piling and allied works. Detailed considerations illustrating the need for undertaking the project are as follows.

(*a*) If standard documents were available considerable time would be saved at the time of tender and misunderstandings would be eliminated. The use of standard forms of contract, such as the ICE Conditions and the RIBA Conditions, is ample confirmation of this point.

(*b*) Many specifications include clauses that are essentially conditions of contract, methods of measurement and instructions to tenderers, whereas a specification should properly be limited to technical matters. In some cases specification clauses duplicate clauses in main and sub-contract conditions, so leading to undesirable legal ambiguity.

(*c*) It is desirable to reduce and, if possible, eliminate the special conditions which specialist piling contractors generally include with their tenders. These conditions usually concern terms of payment, methods of measurement and facilities to be provided and are not necessarily in conflict with the contract conditions for the main works. Although engineers publicly deprecate their use, they are, by and large, admitted. However, the many different special conditions from each tenderer, albeit variations on the same theme, make direct comparison of tenders difficult. The use of a model set of documents, acceptable to all sides of the industry, would assist in this comparison and save time and arguments during construction.

(*d*) There is considerable diversity in the way piling contracts are awarded, ranging from a simple letter asking contractors to quote and a simple letter of acceptance to the most elaborate and formal contract documents. The first extreme is almost invariably inadequate for a contract satisfactory to both parties and the second might be inappropriate for the intended purpose. Model procedures for placing contracts would help avoid the more common pitfalls experienced.

(*e*) A piling specification should command the wide support of the industry because it is in the best interest of the Employer and the public that piling operations should be carried out to standards and in a manner satisfactory to all parties. It would also lead to the elimination of those cases where, unfortunately, piling is undertaken without a specification.

(*f*) The responsibility for the design of piles has tended to become blurred and, in many contracts where designs and guarantees for piling are offered by contractors and accepted by employers, strong doubts have been expressed as to where the ultimate responsibility for the design of the piles lies. The matter of responsibility needs to be clarified. There has been a great increase in the amount of piling work since the 1940s, much of which is undertaken by specialist piling contractors. There has also been a change in the type of piles, precast concrete piles having been largely superseded by cast in situ piles and, for marine work, by steel piles. Large diameter augered piles and piles drilled and formed through bentonite or other drilling fluids have been used increasingly since the 1950s.

(*g*) A specification in accordance with the recommendations for good piling practice given in British Standard CP 2004 but having contractual force should be available. A code of practice can only recommend and cannot be used in place of a specification.

(*h*) A specification for site investigation should be available to ensure adequate and proper exploration of a site before piling takes place. Regrettably in many investigations there are serious shortcomings, some of which have been identified in a NEDO monograph.* Inadequate expenditure on site investigations and a lack of appreciation on all sides of the necessary objectives have had much to do with these shortcomings.

This document contains model procedures for piling, covering contract procedure, conditions of contract, specifications and bills of quantities for piling, a model specification for piling for use in contract documents, and model procedures on site investigation for piling and a model specification for this work. Reference to the model procedures for piling and site investigation for piling must not be made in contracts.

---

*National Economic Development Office. *Action on the Banwell Report*. HMSO, London, 1967.

The model piling contract procedure is presented in the order that is usually found in tender documents. It has been prepared to be consistent with the main forms of contract conditions current in 1977, namely the ICE Conditions (5th edition) and the RIBA Conditions (1963, 1977 revision). In particular the model special clauses for sub-contract are in such a form that they may be used in any piling sub-contract, whether nominated or not. They are given in versions for use with the ICE Conditions and the RIBA Conditions.

Subject to minor variations this document is suitable for use with *General conditions of government contracts for building and civil engineering works.*†

Because of the varied contractual procedures under which piling operations are carried out and the many and varied piling methods and scales of operations, it has not been possible to be definitive about contractual conditions and administrative procedures; nevertheless essential matters are dealt with.

The model specifications (identified by a black line down the edge of the page) are suitable for use with all the main forms of contract, and it is hoped that they will be used throughout the construction industry.

Suggested measured items in bills of quantities are also included in this document. They have been prepared in accordance with the CESMM‡ and those methods of measurement in use in building and civil engineering which have been found to be widely acceptable to all sides. The recommendations are complementary to the CESMM, the RICS–NFBTE SMM§ and the DoE SMM.¶ They are not an alternative to and do not affect the priority of main contract conditions.

Drafting panels of the Piling Specifications Committee were drawn from government departments, public authorities, nationalized industries, consulting engineers, general contractors and specialist piling contractors.

Neither the Institution of Civil Engineers nor the Piling Specifications Committee accepts responsibility for the use which is made of this document. Persons drafting contract documents, specifications and so on are responsible for ensuring that any wording taken from this document is compatible with the remainder of their draft material.

†HMSO, London, edn 1, 1973, Form GC/Works/1.
‡*Civil engineering standard method of measurement*. Institution of Civil Engineers, London, 1976.
§*Standard method of measurement of building works*, 5th edn. Royal Institution of Chartered Surveyors and National Federation of Building Trades Employers, London, 1970.
¶*Method of measurement for road and bridge works*. HMSO, London, 1971.

MODEL PROCEDURES FOR PILING

These model procedures are for guidance only and are to be excluded from contract documents

RESPONSIBILITY FOR PILING

INTRODUCTION

Expertise in piling is concentrated in few hands and work is carried out under varying contractual arrangements, some of which inadequately recognize the technical nature of the work. In addition, because of the varying contractual arrangements, the responsibility for piling is frequently ill defined.

2. As piles are basically structural components, forming part of foundation systems, the rational practice is for a qualified engineer* to be responsible for their design, as he is for the structure they support. In whatever manner the design work is carried out the piling contractor must be responsible for workmanship and his own operations.

CONTRACTUAL ARRANGEMENTS

3. The following are the four basic forms of contractual arrangements which are used for piling work

(a) civil engineering work with an Engineer responsible for design and supervision, either in the employ of or appointed to act on behalf of an Employer

(b) building work with an Architect responsible to an Employer for design and supervision and advised by a civil or structural engineer having no direct contractual responsibility

(c) building work with a Contractor responsible to an Employer for design, supervision and workmanship on behalf of an Employer

(d) building work with an Architect responsible to an Employer for design and supervision but having no independent engineering advice.

TECHNICAL RESPONSIBILITY

4. Major areas of responsibility which require clarification include

(a) *geotechnical information*, which is the principal factor governing design and choice of type of piling for a given structure or loading

(b) *construction operations*, which are affected by the adequacy of the available information concerning the Site and existing piles and structures, and which may affect existing piles and structures and piles constructed under the Contract

(c) *control factors* governing the length of driven piles, their driving time and resistance

---

* A chartered engineer or other engineer having experience and qualifications appropriate for piling work.

(*d*) *proprietary systems*, which depend on the expertise of the contractors offering them, for the provision of satisfactory piles.

ARRANGEMENTS FOR RESPONSIBILITY

5.   The preferred arrangement for responsibility is as follows.

(*a*) Matters concerning soil are of primary consideration. As the Employer looks to the Engineer to advise him on all pertinent matters, the Engineer should be responsible for advising the Employer of the extent of the site investigation required for any proposal and for ensuring that the interpretation of its results is provided; he should also be responsible for ensuring that soil data and reports thereof are preferably included in the tender and contract documents or made available for inspection by the contractors tendering.

(*b*) If the type or method of piling is not specified in the tender documents then the type should be agreed at the time the Contract is placed. Subsequently, decisions affecting the design arising as a result of conditions met on the Site should be the responsibility of the Engineer.

(*c*) Construction operations and workmanship cannot be other than the Contractor's responsibility, although in deciding on his proposed methods when tendering he must necessarily rely on the information made available to him and accessible to him. Tender documents and procedures should define other information provided.

6.   The contractual arrangement set out in paragraph 3(*a*) is in accordance with the proposals for responsibility in paragraph 5.

7.   The arrangement set out in paragraph 3(*b*), although capable of being in accordance with paragraph 5, would be strengthened if the civil or structural engineer were given a direct contractual status *vis-à-vis* the piling contractor by formal delegation of responsibilities from the Architect.

8.   The arrangement in paragraph 3(*c*) can function satisfactorily if the engineering responsibilities are given to a qualified engineer on the Contractor's staff, and can thus be demonstrated to be devolved. The Engineer is then not limited in the exercise of the responsibilities even though the Contractor himself remains legally liable for the fulfilment of the Contract.

9.   The arrangement in paragraph 3(*d*) is fundamentally unsound for engineering works and is to be actively discouraged.

10.   Paragraphs 11–156 are based on the preferred arrangements for responsibility outlined in paragraph 5.

APPOINTMENT OF CONTRACTORS AND SUB-CONTRACTORS FOR PILING

GENERAL

11. Most piling work is carried out by sub-contractors for main contractors working in the civil engineering or building industries under the appropriate form of main contract. Occasionally piling contractors work as main contractors directly for the Employer. The procedures in paragraphs 12–36 apply to particular cases where piling is carried out as a main contract or as a sub-contract under the ICE Conditions or the RIBA Conditions.

PROCEDURES FOR THOSE SEEKING TENDERS

12. Attention is drawn to the section on responsibility for piling (paragraphs 1–10). If the Contractor is required to take responsibility for the design or specification of the permanent piles a clause to that effect shall be included by the Engineer (see, e.g. clause 8(2), ICE Conditions).

13. The Engineer should specify the type of piling which will be billed according to the recommendations for the measurement of piling (paragraphs 99–156), the draft preamble on pages 39–41 and the suggested list of measured items on pages 42–52.

14. The Engineer may consider more than one form of piling suitable and if so alternative bills of quantities should be prepared to be priced according to the piling chosen by the tenderer.

15. The advantages of pre-tender consultation with specialist piling contractors should be considered.

16. It should be stated whether or not consideration will be given to alternative proposals for piling made by a tenderer.

17. If the Contractor is required to insure against direct and consequential loss due to faulty workmanship and materials, the amount of such insurance should be stated in an item in the Bill of Quantities provided in accordance with the conditions of contract (see, e.g., clause 21, paragraph 2, ICE Conditions). Such insurance is not usually required; if it is obtainable it is likely to have many conditional clauses.

18. Any contractor whose tender is not accepted should be so advised as soon as possible.

DOCUMENTS TO BE ISSUED TO TENDERERS

Instructions to tenderers

19. Instructions to tenderers do not form part of the Contract; they should include where appropriate

 (a) the date by which tenders have to be received

 (b) the procedure for submitting tenders

(c) the expected dates of the award of the Contract and of the commencement of the work

(d) any restrictions on visiting the Site and the name, address and telephone number of the person to whom the tenderer should apply for permission to visit the Site should this be necessary

(e) the requirement that tenderers should submit with their tenders a programme for the information of the Engineer

(f) notes on access and any particular site conditions such as adjacent structures

For sub-contracts the following additional item should be given

(g) the name and address of the Main Contractor, if he has been appointed.

Documents which will form part of the Contract

20. Documents which will form part of the Contract are

(a) form of tender

(b) conditions of contract

(c) special conditions of contract required by a particular Employer

(d) special clauses applicable to piling work

(e) Specification

(f) Particular Specification

(g) Drawings

(h) Bill of Quantities

For sub-contracts the following additional items should be included

(i) conditions of sub-contract

(j) special conditions of sub-contract applicable to an individual main contractor.

Information to be included in the tender documents

21. One of the documents that form part of the tender and referred to in paragraph 20 should state

(a) the period for which the tender is to be valid

(b) information regarding damages, retention, insurances (including insurance against direct and consequential loss) and price fluctuations

(c) the location of the Site

(d) the area defined for storage

(e) any conditions concerning limitations on noise and working hours

(f) the number of visits that it will be necessary for the piling contractor to make due to the phasing of the Works and the technical piling considerations

(g) pile cut-off levels

5

(*h*) information on ground and subsoil conditions
(*i*) information on groundwater levels
(*j*) any preliminary pile test results
(*k*) information on existing underground structures and services (preferably in the form of a drawing)
(*l*) the loadings for which the piles have been designed or loadings for which piles have to be designed
(*m*) the class of loadings to which piles may be subjected
(*n*) the test loads, which should be given as

proof test by maintained load (compression)	. . .t
proof test by maintained load (tension)	. . .t
constant rate of penetration (CRP) test	. . .t
constant rate of uplift (CRU) test	. . .t

(*o*) the working level or ground level from which piling will be measured
(*p*) the piling platform level if it is known to be different from the working level.

INFORMATION TO BE SUPPLIED BY THE TENDERER

22. It is preferable that a tender should not be qualified by the tenderer, but the tenderer should state the following and if necessary include any comments thereon in a covering letter with his tender

(*a*) *period tender is open:* if it is not stated in the enquiry, the period for which his tender is open for acceptance, and the period during which work will be done
(*b*) *insurance:* if he is unable to offer insurance for the full amount required, the upper limit of insurance which he can offer
(*c*) *access and site preparation:* the site preparation on which his tender is based if it is different from that given in the tender documents
(*d*) *water:* his requirements for water supply
(*e*) *electric power:* his requirements for electric power
(*f*) *working periods:* the number of visits, the duration of working shifts and the number of hours per week which he proposes to work and on which his tender is based
(*g*) *method statement:* this is particularly important when the piling contractor's work will be followed closely by other operations.

ENQUIRIES FOR SUB-CONTRACT TENDERS

23. Enquiries for sub-contract tenders may be made in the following ways.

(*a*) The Engineer may seek prices directly from piling contractors for nomination as Sub-contractors, prior to, at the same time as or after the main tenders are sought, a Prime Cost Item or a Provisional Sum for the piling being included in the Main Contract. In this event the Engineer must include in the documents for the Sub-contract as much information about the terms of the Main Contract as will be reasonably necessary to ensure that when the Main Contract is awarded the Main Contract and Nominated Sub-contracts will be compatible.

(b) The Engineer may instruct the Main Contractor to invite tenders for the piling Sub-contract. In this case the sub-contract enquiries shall be issued through the Main Contractor who shall first submit his form of sub-contract to the Engineer for approval. Sub-contract tenders shall be returned to the Engineer or Employer either directly or through the Main Contractor as given in the Instructions to Tenderers.

(c) The Engineer may list in the tender documents for the Main Contract the names of approved piling sub-contractors from whom the tenderers for the Main Contract shall seek prices for piling work not undertaken by themselves. The Main Contractor should have the option of proposing further piling contractors for approval. In the absence of an approved list it is customary for the Main Contractor to seek prices directly from his own list of potential sub-contractors. In either case the Sub-contractor is appointed as a direct (domestic) sub-contractor of the Main Contractor, subject to the approval of the Engineer.

PAYMENTS TO THE SUB-CONTRACTOR

24.   Final payment for piling work which is carried out under sub-contract is frequently delayed on large works, as it depends on the release of retention monies under the Main Contract. This should be borne in mind by those arranging contracts and, if it is not possible or reasonable to redress this matter, tenderers should be explicitly informed of the conditions for payment that will be in force.

MAIN CONTRACTOR UNDER THE ICE CONDITIONS

25.   The procedures in paragraphs 12–18, 19(a)–(f), 20(a)–(h), 21 and 22 are applicable.

Model special clauses

26.   The following special clauses are suitable for use in conjunction with the ICE Conditions.

(a) *Obstructions*. If during the execution of the Works the Contractor encounters obstructions in the ground, whether or not these obstructions are referable to clause 12 of the ICE Con-

7

ditions, he shall forthwith notify the Engineer and submit to the Engineer details of the method by which he proposes to overcome the same, and he shall proceed in accordance with the instructions of the Engineer.

(*b*) *Records*. In amplification of clause 56(3) of the ICE Conditions, the signed piling records made in accordance with clause 1.10 of the specification for piling will be used by the Engineer as the basis for measurement.

(*c*) *Improper work and materials*. If in the opinion of the Engineer compliance with clause 39(1) of the ICE Conditions is impracticable the Contractor shall prepare calculations, designs and drawings and supply materials and execute work to provide a foundation acceptable to the Engineer.

SUB-CONTRACTOR UNDER THE ICE CONDITIONS

27. The procedures given in paragraphs 12–24 are applicable.

28. If the piling Sub-contractor is to be nominated by the Engineer in accordance with clause 59 of the ICE Conditions the procedure given in paragraphs 23(*a*) and 23(*b*) is appropriate. However, if the piling Sub-contractor is to be appointed by the Main Contractor on his own responsibility paragraph 23(*c*) is appropriate as a guide to him in seeking prices for the work. In this event and for the procedures in paragraphs 12–22 it will be necessary to read 'Contractor' for 'Engineer' as appropriate.

29. If the Sub-contractor is required to take responsibility for design and specification of piling then a special clause for clause 58(3) of the ICE Conditions will be required to state expressly the obligations of the Main Contractor.

Model special clauses

30. The following special clauses are suitable for use in conjunction with the ICE Conditions and the FCEC Form of Sub-contract.

(*a*) *Ground and subsoil conditions*. The Contractor shall make available to the Sub-contractor all relevant information provided by or on behalf of the Employer pursuant to clause 11 of the ICE Conditions, including that relating to the nature of the ground and subsoil, together with any information that has been provided to him regarding the location, depth and condition of underground services which may be affected by the piling work. The Contractor shall also make available to the Sub-contractor any available additional information relevant to the piling work he has obtained.

(*b*) *Obstructions*. If during the execution of the Works the Sub-contractor encounters obstructions in the ground, whether or not these obstructions are referable to clause 12 of the ICE Conditions, he shall forthwith notify the Contractor and submit to the Contractor details of the method by which he proposes to overcome the same, and he shall proceed in accordance with the instructions of the Contractor. The Sub-contractor shall, in support of any payment or claim thereto to which he may be entitled under the Sub-contract, give notices, keep records and take such other action as provided for in clause 52(4) of the ICE Conditions as instructed by the Contractor.

(*c*) *Underground structures and services*. If the Sub-contractor damages any underground structures or services, whether or not he has been informed of their presence, he shall forthwith inform the Contractor.

(*d*) *Records*. In amplification of clause 56(3) of the ICE Conditions, the signed piling records made in accordance with clause 1.10 of the specification for piling will be used by the Engineer as the basis for measurement.

(*e*) *Facilities*. The following facilities and services shall be provided by the Main Contractor free of cost to the Sub-contractor under the fourth schedule, part B of the FCEC Form of Sub-contract

(i) negotiations for the procurement of any necessary licences, sanctions or authorities, including any way-leaves, possessions, rights of way or access

(ii) hoardings, fences, watching, traffic control, flagmen or the like as are necessary to protect the Works, plant, materials and personnel of the Sub-contractor

(iii) means of access to reasonably level working surfaces and their maintenance firm enough for the movement on to and off the Site and between pile positions of mobile plant and equipment and vehicles carrying plant and mobile concrete mixers

(iv) a firm and reasonably level area and its maintenance, situated conveniently to the piling area and adequate for storage and preparatory operations, and other site facilities adequate for the requirements given by the Sub-contractor in his tender

(v) any pumping or drainage required to keep the Site free of surface water

(vi) removal of known artificial obstructions which may impede piling operations and backfilling with material which will not obstruct or be deleterious to the piling

(vii) within the piling and the storage and preparatory operations area: mains water take-off points and water for piling operations, concrete mixing and cleaning of plant, and electric power take-off points; these shall be adequate for the requirements given by the Sub-contractor in his tender

(viii) general site lighting

(ix) facilities under the Health and Safety at Work Act 1974 and any amendments or re-enactment thereof.

(*f*) *Setting out.* The Contractor will provide and maintain on the Site permanent datum level points, base lines and grid lines, together with dimensioned drawings from which the pile positions are to be set out by the Sub-contractor. The Sub-contractor shall mark each pile position with a peg. No pile shall be installed until its setting out has been checked by the Contractor. The Sub-contractor shall give adequate notice to the Contractor when he requires setting out to be checked and the Contractor shall carry out such checks so as to avoid delay to the Sub-contractor. Nevertheless the Sub-contractor shall be responsible for the correct and proper setting out of the piles and for the correctness of the positions, levels, dimensions and alignment of the piles, and for the provision of all necessary associated instruments, appliances and labour.

(*g*) *Electric power.* The Sub-contractor shall pay to the Contractor such sums as may be reasonable for the electric power consumed as a result of his operations, other than for general site lighting, unless such lighting is required for the Sub-contractor only. The Sub-contractor shall be responsible for the provision and maintenance of the electrical installation on the load side of the points of supply.

(*h*) *Stripping pile heads.* Unless otherwise directed, the Contractor shall strip the pile heads to cut-off level, straighten the reinforcement and remove surplus material.

(*i*) *Improper work and materials.* If in the opinion of the Engineer compliance with clause 39(1) of the ICE Conditions is impracticable, the Sub-contractor shall prepare calculations, designs and drawings and supply materials and execute work to provide a foundation acceptable to the Contractor and the Engineer.

MAIN CONTRACTOR UNDER THE RIBA CONDITIONS

31.   References to the Engineer will have to be amended to take account of the contractual arrangements for the work which could be an arrangement of paragraph 3(*b*) or 3(*c*).

32.   The procedures in paragraphs 12–18, 19(*a*)–(*f*), 20(*a*)–(*h*), 21 and 22 are applicable.

Model special clauses

33.   The following model special clauses are suitable for use in conjunction with the RIBA Conditions.

(*a*) *Extent of Contract.* The Contract shall, unless otherwise stated and subject to the provisions of the Contract, be deemed to include the provision of all labour, materials, construction plant, Temporary Works, transport to, from, on or about the Site, and everything whether of a temporary or permanent nature required in and for the construction, completion and maintenance of the Works so far as the necessity for providing the same is specified in or may reasonably be inferred from the Contract.

(*b*) *Inspection of Site.* The Contractor shall be deemed to have inspected and examined the Site and its surroundings and to have satisfied himself before submitting his tender as to the nature of the ground and subsoil (so far as is practicable and having taken into account any information in connection therewith which may have been provided by or on behalf of the Employer), the form and nature of the Site, the extent and nature of the work and materials necessary for the completion of the Works, the means of communication with and access to the Site, the accommodation he may require, and in general to have obtained for himself all necessary information (subject to the above-mentioned) as to risks, contingencies and all other circumstances influencing or affecting his tender. With regard to the subsoil conditions the tender shall be deemed to have been based on such information, including the location, depth and condition of underground structures and services, which may be affected by the piling work and which shall have been supplied to the Contractor at the time of tendering.

(*c*) *Obstructions.* If during the execution of the Works the Contractor encounters obstructions in the ground he shall forthwith notify the Architect and submit to the Architect details of the method by which he proposes to overcome the same, and he shall proceed in accordance with the instructions of the Architect. The Contractor shall also submit records as may be reasonably required to support any claim for payment in accordance with the Contract.

(*d*) *Works to be measured.* The quantities set out in the Bill of Quantities are the estimated quantities and are not to be taken as the actual and correct quantities of the work to be carried out by the Contractor in the fulfilment of his obligations under the Contract. The value of the work actually carried out shall be ascertained and determined by measurement of the work actually carried out and application to all items arising therefrom of the rates in the Bill of Quantities or such other rates as shall be determined in accordance with the provisions of the Contract.

(*e*) *Records.* The signed piling records made in accordance with clause 1.10 of the specification for piling will be used by the Architect as the basis for measurement.

(*f*) *Variation in conditions.* In extension of clause 11 of the RIBA Conditions, if in the execution of the Works the Contractor is required to carry out work under conditions substantially different from those given in the documents on which his tender was based, he shall notify the Architect in writing to that effect and that he may claim a variation. In the latter event the Contractor shall supply to the Architect as soon as is practicable an estimate of the extent of any delay and of the additional payment to which he considers himself entitled.

(*g*) *Setting out.* In extension of clause 5 of the RIBA Conditions, the Contractor shall provide and maintain on the Site permanent datum level points, base lines and grid lines, together with dimensioned drawings from which he shall set out the pile positions and mark each pile position with a peg. No pile shall be installed until its setting out has been checked by the Architect; nevertheless the Contractor shall be responsible for the correct and proper setting out of the Works and for the correctness of the positions, levels, dimensions and alignment of all parts of the Works, and for the provision of all necessary associated instruments, appliances and labour.

(*h*) *Improper work and materials.* In extension of clause 6 of the RIBA Conditions, in the event of the removal of any work, materials or goods which are not in accordance with the Contract being impracticable, the Contractor shall be responsible for the costs of all additional work, including the costs of any calculations, designs and drawings which may become necessary to provide a foundation acceptable to the Architect.

(*i*) *Clearance of Site on completion.* The Contractor shall clear away and remove from the Site all rubbish, debris and spoil as it accumulates and on completion of his work leave all areas occupied by him clean and in a workmanlike condition and to the satisfaction of the Architect.

SUB-CONTRACTOR UNDER THE RIBA CONDITIONS

34.   It is assumed in paragraphs 35–36 that the piling contractor will be the Nominated Sub-contractor or a direct Sub-contractor of the Main Contractor. References to the Engineer will have to be amended to take account of the contractual arrangements for the work which could be an arrangement of paragraph 3(*b*) or 3(*c*).

35.   The procedures in paragraphs 12–24 are applicable.

Model special clauses

36.   The following model special clauses are suitable for use in conjunction with either (i), (ii) and (iii) or (i) and (iv)

(i) the *Standard form* of building contract, 1963, 1977 revision, published for the Joint Contracts Tribunal by Royal Institute of

British Architects Publications Ltd, issued by the Joint Contracts Tribunal, herein referred to as the RIBA Conditions

(ii) the form of sub-contract *For use where the Sub-contractor is nominated under the Standard form of building contract issued by the Joint Contracts Tribunal (1963 edition as revised)*, 1963, 1977 revision, issued under the sanction of and approved by the National Federation of Building Trades Employers and the Federation of Associations of Specialists and Sub-contractors and approved by the Committee of Associations of Specialist Engineering Contractors, herein referred to as the Green Form

(iii) the *Standard form of tender for Nominated Sub-contractors*, published for the Joint Contracts Tribunal by Royal Institute of British Architects Publications Ltd

(iv) the form of sub-contract *For use where the Sub-contractor is not nominated by the Architect*, 1971, 1977 revision, approved by the National Federation of Building Trades Employers, the Federation of Associations of Specialists and Sub-contractors and the Committee of Associations of Specialist Engineering Contractors.

(*a*) *Extent of Contract*. The Sub-contract shall, unless otherwise stated and subject to the provisions of the Sub-contract, be deemed to include the provision of all labour, materials, construction plant, Temporary Works, transport to, from, on or about the Site, and everything whether of a temporary or permanent nature required in and for the construction, completion and maintenance of the Works so far as the necessity for providing the same is specified in or may reasonably to be inferred from the Sub-contract.

(*b*) *Inspection of Site*. The Sub-contractor shall be deemed to have inspected and examined the Site and its surroundings and to have satisfied himself before submitting his tender as to the nature of the ground and subsoil (so far as is practicable and having taken into account any information in connection therewith which may have been provided by or on behalf of the Contractor), the form and nature of the Site, the extent and nature of the work and materials necessary for the completion of the Works, the means of communication with and access to the Site, the accommodation he may require, and in general to have obtained for himself all necessary information (subject to the above-mentioned) as to risks, contingencies and all other circumstances influencing or affecting his tender. With regard to subsoil conditions the tender shall be deemed to have been based on such information, including the location, depth and condition of underground structures and services, which may be affected by the piling work and

13

which shall have been supplied to the Sub-contractor at the time of tendering.

(*c*) *Obstructions.* If during the execution of the Works the Sub-contractor encounters obstructions in the ground he shall forthwith notify the Contractor and submit to the Contractor details of the method by which he proposes to overcome the same, and he shall proceed in accordance with the instructions of the Contractor. The Sub-conctractor shall also submit records as may be reasonably required to support any claim for payment in accordance with the Sub-contract.

(*d*) *Underground structures and services.* If the Sub-contractor damages any underground structures or services, whether or not he has been informed of their presence, he shall forthwith inform the Contractor.

(*e*) *Works to be measured.* The quantities set out in the Bill of Quantities are the estimated quantities and are not to be taken as the actual and correct quantities of the work to be carried out by the Sub-contractor in the fulfilment of his obligations under the Sub-contract. The value of the work actually carried out shall be ascertained and determined by measurement of the work actually carried out and application to all items arising therefrom of the rates in the Bill of Quantities or such other rates as shall be determined in accordance with the provisions of the Sub-contract.

(*f*) *Records.* The signed piling records made in accordance with clause 1.10 of the specification for piling will be used by the Contractor as the basis for measurement.

(*g*) *Variation in conditions.* In extension of clause 11 of the RIBA Conditions and clauses 7 and 10 of the Green Form, if in the execution of the Works the Sub-contractor is required to carry out work under conditions substantially different from those given in the documents on which his tender was based, he shall notify the Contractor in writing to that effect and that he may claim a variation. In the latter event the Sub-contractor shall supply to the Contractor as soon as is practicable an estimate of the extent of any delay and of the additional payment to which he considers himself entitled.

(*h*) *Facilities for Sub-contractor.* The Contractor shall provide the Sub-contractor free of charge the facilities and services defined in the Standard Form of Tender for Nominated Sub-contractors and in addition the following facilities and services

(i) negotiations for the procurement of any necessary licences, sanctions or authorities, including any way-leaves, possessions, rights of way or access

(ii) hoardings, fences, watching, traffic control, flagmen or the like as are necessary to protect the Works, plant, materials and personnel of the Sub-contractor

(iii) means of access to reasonably level working surfaces and their maintenance firm enough for the movement on to and off the Site and between pile positions of mobile plant and equipment and vehicles carrying plant and mobile concrete mixers

(iv) a firm and reasonably level area and its maintenance, situated conveniently to the piling area and adequate for storage and preparatory operations, and other site facilities adequate for the requirements given by the Sub-contractor in his tender

(v) any pumping or drainage required to keep the Site free of surface water

(vi) removal of known artificial obstructions which may impede piling operations and backfilling with material which will not obstruct or be deleterious to the piling

(vii) within the piling area and the storage and preparatory operations area: mains water take-off points and water for piling operations, concrete mixing and cleaning of plant, and electric power take-off points; these shall be adequate for the requirements given by the Sub-contractor in his tender

(viii) general site lighting

(ix) facilities under the Health and Safety at Work Act 1974 and any amendments or re-enactment thereof.

(*i*) *Setting out*. In extension of clause 5 of the RIBA Conditions, the Contractor will provide and maintain on the Site permanent datum level points, base lines and grid lines, together with dimensioned drawings from which the pile positions are to be set out by the Sub-contractor. The Sub-contractor shall mark each pile position with a peg. No pile shall be installed until its setting out has been checked by the Architect and/or the Contractor. Nevertheless the Sub-contractor shall be responsible for the correct and proper setting out of the piles and for the correctness of the positions, levels, dimensions and alignment of the piles, and for the provision of all necessary associated instruments, appliances and labour.

(*j*) *Electric power*. The Sub-contractor shall pay to the Contractor such sums as may be reasonable for the electric power consumed as a result of his operations, other than for the general site lighting, unless such lighting is required for the Sub-contractor only. The Sub-contractor shall be responsible for the provision and maintenance of the electrical installation on the load side of the points of supply.

15

(*k*) *Stripping pile heads.* Unless otherwise directed, the Contractor shall strip the pile heads to cut-off level, straighten the reinforcement and remove surplus material.

(*l*) *Improper work and materials.* In extension of clause 6 of the RIBA Conditions, in the event of the removal of any work, materials or goods which are not in accordance with the Subcontract being impracticable, the Sub-contractor shall be responsible for the costs of all additional work, including the costs of any calculations, designs and drawings which may become necessary to provide a foundation acceptable to the Contractor.

(*m*) *Clearance of Site on completion.* The Sub-contractor shall clear away and remove from the Site all rubbish, debris and spoil as it accumulates and on completion of his work leave all areas occupied by him clean and in a workmanlike condition and to the satisfaction of the Contractor.

USE OF THE MODEL SPECIFICATION FOR PILING

GENERAL

37. The model specification for piling (pages 53–121) applies to bearing piles only and is consistent with CP 2004.

38. The Engineer may use the model specification in three ways

(*a*) by including in the contract documents the whole or complete sections of it, together with a statement of amendments and the Particular Specification

(*b*) by referring in the contract documents to the whole or complete sections of it, and including a statement of amendments and the Particular Specification

(*c*) by using it as a basis for his own Specification.

Method (*c*) should not be used unless the Contract is one of exceptional magnitude or complexity.

39. Every effort has been made to avoid conflict between the model specification and the ICE Conditions and the RIBA Conditions. However, certain clauses in the ICE Conditions do not have parallel clauses in the RIBA Conditions. Therefore clauses in the model specification which the Engineer considers covered by the ICE Conditions or the RIBA Conditions and/or the Specification for the Main Contract (if appropriate) must be amended or deleted by the Engineer.

40. Where a given type of pile is to be specified only the appropriate sections need be used, e.g. for driven cast in place piles sections 1, 2, 5 and 8. However, in many instances it may be felt advantageous to have the benefit of the model specification as a whole, so that the Contract will cover any unexpected developments that may lead to a change in the type of pile.* With this arrangement there are bound to be clauses repeated throughout the Specification, but it does allow each section to be complete in itself without excessive cross-referencing. Where a clause contains more than one standard for material or workmanship then a clause will be required in the Particular Specification to define which standard is to apply.

GENERAL REQUIREMENTS FOR PILING WORK

41. A number of general clauses which are desirable for any contract are listed in section 1 of the model specification. This is not intended to be a comprehensive list because many of those clauses and similar matters might be covered by the Employer's or Engineer's normal Specification. Clause 1.021 (*e*) (working area) may require

---

* Weltman, A. J. and Little, J. A. *A review of bearing pile types.* Property Services Agency and Construction Industry Research and Information Association, London, 1977.

17

amplification for contracts under the ICE Conditions as follows: 'the allocation of working areas to the Contractor does not afford him exclusive rights of occupation; the Employer and other contractors shall have the right to complete freedom of access to all parts of the Works and to carry out any work that may be required in any place on the Site.' Before including such amplification for works carried out under the RIBA Conditions and other conditions it is desirable to check whether or not it conflicts with the Main Contract (see particularly clause 21(1) of the RIBA Conditions). Legal advice may be needed.

42.   Additional clauses would be required, for instance, in the case of piling for maritime work because the Contractor would require the right to use a berth for his craft; the Contractor would have to know whether port and wharf dues were payable by him.

PARTICULAR SPECIFICATION CLAUSES

43.   Sections 2–8 of the model specification require clauses to be given in the Particular Specification to cover items such as the class of cement, concrete, strength requirements and minimum penetration. The clauses of the Particular Specification should be prepared and set out in the order of the sections and clauses of the model specification.

LIMITS OF THE MODEL SPECIFICATION

44.   The model specification is a specification of materials and workmanship, and does not attempt to specify design criteria.

RESPONSIBILITY FOR DESIGN OF PILING

45.   The large amount of piling work carried out which is contractor-designed should be viewed in the light of the recommendations in paragraphs 1–10; it is referred to in clause 1.08. Where contractor-designed piling is to be used special attention must be given to the definition of design criteria and responsibilities; these are not covered by the model specification.

USE WITH FORMS OF CONTRACT

46.   The model specification may be used with all the main forms of contract and with nominated or domestic sub-contracts but it will be necessary in some cases to amend the words 'Engineer' and 'Contractor' in prefatory notes. For example, in nominated sub-contract works carried out under the RIBA Conditions a note should be included, 'For "Contractor" read "Sub-contractor". In other contexts, for "Engineer" read "Architect" or "Supervising Officer".'

ENGINEER'S REPRESENTATIVE

47. In this document reference is made only to the Engineer. In the event of an Engineer's Representative being appointed the Engineer should inform the Contractor in writing of any powers which he may delegate to the Engineer's Representative.

RECORDS

48. The signed records will form a record of the work (clause 1.10). In most contracts the Engineer has to prepare 'as-made' drawings of the work on which he would normally summarize the information from the daily piling records. These drawings also have to include the positions of obstructions, piles abandoned and rejected and the results of pile tests. The importance of these records and as-made drawings cannot be over-emphasized.

COMPLIANCE WITH CP 110

49. Section 2 of the model specification assumes compliance with CP 110.* Problems can arise in applying CP 110 to works below ground where the factoring of active and passive soil pressures and groundwater pressures is not yet wholly understood. For this reason CP 110 and CE CP 2† are not compatible‡ and attempts to use them in combination can be hazardous. Likewise many piles require analysis for bending due to the conditions of loading and the same problems arise.

50. The basis for judging the quality of concrete and characteristic strength in CP 110 is statistical. On many piling sites the volumes of concrete used are relatively small, and the duration of works is short. The treatment of strength on a statistical basis is scarcely justified in some cases. Moreover, the minimum cement contents given in Table 48 of CP 110 are not sufficient for cast in place piles, where criteria other than strength, such as high workability and coherence, are primary requirements. The higher cement contents given in the model specification will normally be sufficient to give a more than adequate reserve of strength.

51. For these reasons specification in accordance with CP 110 in cast in place piling is by no means universal at present; many engineers prefer to specify in accordance with CP 114.§ The specification writer should carefully consider these points and make any necessary amendments to the model specification in accordance with his requirements.

---

*The structural use of concrete. British Standards Institution, London, 1972, CP 110.
†Earth retaining structures. British Standards Institution, London, 1951, CE CP 2.
‡Design and construction of deep basements. Institution of Structural Engineers, London, 1975.
§Structural use of reinforced concrete in buildings. British Standards Institution, London, 1969, CP 114.

AGGREGATES

52. In clause 2.064 it is stated that Table 3 is based on the use of gravel aggregates and is applicable to concrete for use in cast in place piling. This does not mean that crushed stone cannot be used for cast in place piling.

BENTONITE*

53. The use of bentonite fluid is referred to in section 4 of the model specification. The same general principles would apply to the use of any other type of drilling fluid but clauses concerning the properties of the fluid would have to be amended by means of clauses in the Particular Specification.

SETTING OUT OF CAST IN PLACE PILES

54. Reference is made in clause 4.06 to checks on the position of a pile casing during and immediately after placing. These are essential in those areas where overbreak occurs, because once the pile has been concreted the apparent centre of the top may not be concentric with the pile shaft, the position of which was checked previously.

DRIVING RESISTANCE AND PENETRATION

55. Clauses shall be given in the Particular Specification to cover driving resistance, penetration and designed loads. The use and wording of such clauses will depend on the type of contract and the possibility that a minimum or a maximum penetration is a feature of the design.

STEEL PILES

56. The model specification does not refer to steel sheet piling, of which there are two basic types: a trough or Z type and the straight web type. However, many of the clauses concerning the fabrication. coating and driving of steel piles may be used in the preparation of clauses for steel sheet piling.

57. The model specification has been prepared for piles that are mainly load bearing piles. This broad classification includes

(*a*) top driven piles, sometimes hearted with concrete and normally under static compressive or tensile loads

(*b*) bottom driven piles, sometimes hearted with concrete and normally under static compressive or tensile loads but subject also to tensile loads during driving

(*c*) free standing piles, often of large diameter and thickness that

*Fleming, W. G. K. and Sliwinski, Z. J. *The use and influence of bentonite in bored pile construction*. Property Services Agency and Construction Industry Research and Information Association, London, 1977.

are usually top driven; these are normally under a static compressive or tensile load but sometimes they are free standing and receive impact loads from the berthing of ships that induce fluctuations in stress; they are often subject to loads such as those from wind, waves and currents that induce reversal of stresses.

58.   Clause 6.14 covers the acceptance of radiographs of welds for piles subject to different classes of loading. The class of loading to which the piles may be subjected must be stated in the Specification.

59.   In the case of piles that may be subject to reversal of stresses the designers must consider whether a relaxation of the standards of circumferential weld reinforcement is permissible and specify such in clause 6.03(*d*). In addition the classes of loads other than driving and vertical loads to which a pile may be subjected must be given in clause 1.021(*b*).

PILE TESTING

60.　Most piles when constructed or installed are not accessible for inspection and so it is desirable that in all but the simple and straightforward cases some piles should be subject to testing before being incorporated in structures.

61.　General recommendations on the interpretation and application of pile test results with regard to the use of piles in various structural conditions, are given in CP 2004.* Other tests applicable to piles have been considered but their development and interpretation requires further study and so specifications for them cannot at present be written for general application. However, these tests can be of value and it is hoped that their use will become more common so that a standard practice will become established.

62.　Tests available include

(*a*) determination of integrity

(*b*) check of straightness

(*c*) lateral load testing

(*d*) direct load testing along the pile axis.

63.　Comments on these tests are now given. Detailed comments are confined to (*d*) as it is the only test in common use and it is generally considered the important criterion.

DETERMINATION OF INTEGRITY

64.　The tests which determine integrity are generally applied to cast in place piles, particularly to large diameter bored piles, although certain tests are more suitable for precast piles. Reference should be made to *Integrity testing of piles: a review.*†

65.　Cast in place piles may contain voids, breaks or waists because of the use of concrete unsuitable for the method of placement, inflow of the soil or water into the borehole, localized high soil pressures, or lifting of the concrete and separation of the pile shaft as the casing tube is withdrawn.‡ Various methods have been devised for giving an indication of the presence of defects, some of which have been in commercial operation for several years. Certain of these test methods may be applied to the pile after completion; for others the decision to test has to be made before construction so that tubes allowing access for

---

*Foundations. British Standards Institution, London, 1972, CP 2004.

†Weltman, A. J. *Integrity testing of piles: a review.* Property Services Agency and Construction Industry Research and Information Association, London, 1977.

‡Thorburn, S. and Thorburn, J. Q. *Review of problems associated with the construction of cast-in-place concrete piles.* Property Services Agency and Construction Industry Research and Information Association, London, 1977.

the test equipment down the length of the pile may be provided and cast into the concrete.

66. A borehole drilled or cored down the length of the pile will indicate the presence of voids on the line of the bore. Examination of the sides of the borehole by a closed circuit television camera will show gaps or cracks and the quality of the concrete through which the bore passes. Vision is impaired if the hole fills with cloudy water. A flocculating agent may be used to clarify the water. Examination of the profile of the bore with borehole calipers or by acoustic means will indicate any zones of overbreak in the bore, absence of concrete and, to some degree, the quality of the concrete.

67. The position of a borehole drilled in a pile shaft relative to the pile axis may become uncertain as the depth from the surface increases, so that an accurate location of any defect within the pile cross-section is unlikely.

68. If the borehole is formed by coring then inspection of the core extracted will give an indication of the concrete quality. The borehole only enables identification of the material which is brought out or which may be observed at the sides of the hole. The nature of the remainder of the pile shaft has to be deduced from the circumstances.

69. By using a proprietary acoustic logging apparatus in which a transmitter of sonic pulses and a receiver, separated by an acoustic insulator, are lowered down a hole in the pile, an indication of the quality of the concrete within about 100 mm around the hole is obtained. The hole may be drilled or a tube may be cast in the pile. By using a method based on the transmission of sonic pulses through the material between a pair of parallel tubes cast in the pile, defects lying between the tubes are revealed. The method, called twin tube logging, has been in regular use for some time. A limitation is that vertical features of 75 mm or less may be missed. This dimension may be reduced if a second scan is carried out with the probes at slightly different depths relative to each other. The tubes must be adequately fixed to maintain their position while concrete is placed.

70. Testing methods based on neutron or gamma radiation are available. Generally a back-scattering technique is used, in which the radiation source and receiver, separated by a shield, are lowered into a hole, which may be either bored or preformed by casting in a tube. Neutron scatter gives an indication of the moisture content and gamma ray scatter an indication of the density of the material within about 100 mm around the hole. Alternatively, radiation may be transmitted between a pair of tubes cast in the pile during construction, although a more powerful source is usually necessary with this technique. The back-scattering method has been in regular use for

some time. Sufficient tubes are cast in the pile to permit the desired cross-section to be scanned.

71.   In the vibration method of testing an electro-dynamic vibrator and a velocity transducer are mounted on the pile head. The pile is vibrated at frequencies of 1–1000 Hz. The signal from the velocity transducer is a measure of the mechanical admittance of the pile head and is plotted on an $X-Y$ recorder against frequency as the test proceeds. Peaks of resonance occur in the plot, their position being determined by the distance from the pile base or a discontinuity or change in shaft section.

72.   In impact or seismic methods the vibrational response of the pile to a blow on the head is recorded by means of an oscilloscope. The record is analysed to obtain the length of pile brought into vibration, i.e. the depth to the base or a discontinuity. A routine method has been developed in the Netherlands for testing reinforced concrete driven piles.

73.   Acoustic and radiation logging, and vibration, seismic and impact methods all need special apparatus, which is usually operated by a proprietary firm; the field results require interpretation by a specialist with experience of the method adopted. However, it remains the Engineer's responsibility to decide whether or not any particular defect which is reported is of sufficient significance to justify further investigation or remedial action.

74.   A method of testing that applies the design compressive load to the pile shaft has been the subject of experiment and involves the use of sleeved rods or looped cables, anchored in the base, which are cast into the pile such that by jacking between the pile head and the rods a compressive load is applied to the pile shaft. The stressing rods or cables may be recovered after the test.

CHECK OF STRAIGHTNESS

75.   Tests for straightness, vertically or at a specific inclination, involve, in the case of cast in place piles, the measurement of the alignment of a borehole or tube. In the case of preformed driven piles, before driving to permit the tests to be made, the attachment or embedment of a duct is required. In neither case is it considered that use of the tests is sufficiently common to warrant inclusion in the model specification.

LATERAL LOAD TESTING

76.   Lateral loading tests on piles have been made infrequently and there is little general experience of the methods for such tests. What can be done in a given case is considerably influenced by the local conditions and circumstances so that an engineer requiring a lateral

load test must specify a procedure having regard to the interpretation of the test and its relevance to structural design. Guidance on lateral pile testing has been given by Wagner.*

77.   It is possible to overstress the pile in this test, particularly if the pile has a free standing length.

DIRECT LOAD TESTING ALONG THE PILE AXIS

78.   Load testing establishes the capability of the pile to carry the test load, but it gives no other indication of its compliance with the Specification.

79.   Load testing may be carried out by

(*a*) maintained load testing

(*b*) the CRP test and the CRU test.

The relative merits of methods (*a*) and (*b*) have been presented by Whitaker and Cooke.†

Number of piles to be tested

80.   It is advisable to test at least one preliminary pile for each major grouping of piles on a site or at least one preliminary pile for each hundred working piles on a large site. Subsequently, working piles should be tested in such numbers and at locations as will provide reasonable assurance that the Specification is being met and the preliminary test pile performance is being maintained over the whole of the Site.

81.   In deciding the number of working piles to be tested it will be necessary to consider

(*a*) the requirements of the local statutory authority

(*b*) the extent and reliability of the site investigation

(*c*) variations in ground conditions

(*d*) knowledge of the behaviour of similar piles in similar circumstances on adjacent sites

(*e*) the standard of labour and supervision on the Site

(*f*) any unusual difficulty or major variation encountered in the formation of the working piles.

Measurement of load

82.   The load measuring device may consist of a proving ring, load measuring column, pressure cell, strain gauges or other appropriate system.

*Wagner, A. A. Lateral load tests on piles for design information. *Symposium on lateral load tests on piles*. American Society for Testing and Materials, Philadelphia, Special Tech. Pub. 154, 59–72.

†(a) Whitaker, T. and Cooke, R. W. A new approach to pile testing. *Proc. 5th Int. Conf. Soil Mech., Paris*, 1961, **2**, 171–176. (b) Whitaker T. The constant rate of penetration test for the determination of the ultimate bearing capacity of a pile. *Proc. Instn Civ. Engrs*, 1963, **26**, Sept., 119–124. (c) Whitaker, T. *Experiences with the constant rate of penetration test for piles*. Building Research Establishment, Garston, 1967, Engineering series CP 43.

83.   The movement of the pile head should be measured in relation to a frame which is unaffected by the movement of the pile itself, the surrounding ground and the test load. However, the creation of such a reference is often difficult. Movement may be caused by any of the following.

(*a*) Ground displacements due to changes of loading in the vicinity. For example, the vertical displacement of a reference frame, supported on shallow foundations on soft clay as the kentledge load is removed from the ground surface during a test, can be 20 mm or more.

(*b*) Ground displacements due to climatic effects. Movement of the ground surface due to wetting by a shower of rain and subsequent drying is generally negligible, but there would be a slow movement of a foundation placed at shallow depth in a clay soil in a prolonged period of wet or dry weather, e.g. at 30 cm depth in a soft clay the movement could be 10 mm in a month.

(*c*) The effects of temperature and wind on the datum frame structure. The direct rays of the sun cause differential expansion in the frame and consequent warping, and wind causes vibration. A frame which is shielded from sun and wind by boxing will still be subject to daily variation in temperature which may cause a cyclic displacement of 0·25–0·5 mm.

84.   Cyclic movements and their pattern due to causes as in (*c*) can be established if a series of readings is taken before the loading test starts. The effects of cyclic movements may be minimized in a test of long duration if readings are taken at the same time each day.

85.   When constructing the reference frame and its foundations the Engineer must consider the type of test and the precision required in the measurement of movements. In general the results of the CRP or CRU tests are little affected by the foregoing influences.

86.   If the precision required in the maintained load test is greater than that obtainable with the levelling method normally available to the Engineer, then an optical level incorporating a parallel plate micrometer should be used. This enables an accuracy of 0·1 mm to be obtained.

87.   If a reference frame is used, an improvement in the accuracy may be obtained if the frame is supported on piles, or if deep datum points are formed at each foundation in order to measure the movement of the frame relative to a stable stratum.* The Engineer should

---

*Whitaker, T. Discussion on Southern Outfall Works of the London County Council, by E. H. Vick *et al. Proc. Instn Civ. Engrs*, 1966, **33**, Apr., 719–723.

specify the form of optical levelling and the precision required or the particular type of support to be provided for the reference frame.

Method of applying load

88.   The load should be applied by one of the following methods
(*a*) jacking against kentledge adequately supported clear of the pile under test
(*b*) jacking against a frame deriving its reaction from piles clear of the influence of the pile under test
(*c*) jacking against a sufficiently adequate structure.
The direct application of kentledge to a pile is not recommended.

Maintained test load

89.   The Engineer shall state in the contract documents the required maximum test load. When deciding on this value he should take into account the possible variations in the soil strata, which may not have been revealed in the site investigation and which may give rise to a larger ultimate bearing capacity than that estimated. Provided the structural integrity or stability of the pile is not likely to be damaged and the increase in the proposed pile test load is within economic and practical limits, consideration should also be given to making the test load of such value that soil failure should occur. Such a test load will give a guide to the true factor of safety and may assist in the interpretation of variations in the driving characteristics of the working piles.

CRP test or CRU test

90.   In a CRP test on an end bearing pile difficulty is often experienced in identifying the point on the force–penetration diagram at which ultimate bearing capacity is reached. Ultimate bearing capacity of the base is not normally reached until the penetration is equal to at least 10% of the base diameter. At smaller penetrations only a portion of the ultimate bearing capacity is mobilized. Often a penetration equal to 20% or more of the base diameter is required in frictional soils to obtain full mobilization. The figure for any given circumstance is dependent on the state of compaction of the soil. The value of 10% represents the lower limit for full mobilization of the base resistance. For piles with large bases the need for economy in testing generally precludes the achievement of a large penetration unless there is a special need for the information and, on grounds of expediency, the lower limit of penetration at which full mobilization may be expected to occur (i.e. a penetration equal to 10% of the base diameter) has been included in the Specification.
91.   The penetration to be achieved is an arbitrary choice and an engineer wishing to make a closer approach to the value of ultimate

bearing capacity by means of the CRP test should consult the literature.*

92. The rate of penetration or uplift to be used in a particular test should be chosen with regard to the nature of the support which the pile obtains, the elastic compression or extension of the pile, the time required to take and record readings and the number of readings required to plot an unambiguous curve of load against movement. The pump supplying the hydraulic jack must be capable of maintaining the specified rate. The movement of the base of the pile will be smaller than that of the head because of elastic shortening or extension of a pile under load. The elastic shortening or extension may be estimated or may be measured; the method used shall be approved by the Engineer.

93. If the pile is end bearing on rock or very dense gravel a short stiff pile may reach the maximum specified test load with a very small penetration of the base and consequently little downward movement of the head. A long pile, although having a small base penetration, may show an appreciable downward movement of the head due to elastic compression. An estimate must be made of this downward movement so that sufficient readings may be obtained to cover the range from zero to the maximum test load.

94. If the pile is end bearing in medium to dense sand or gravel a rate of penetration of 1·5 mm/min is generally convenient for piles up to 20 m long. For a very long pile this rate may be increased to allow for the elastic compression of the pile.

95. If the pile obtains the greater part of its support by skin friction on the pile shaft the movement to reach ultimate bearing capacity may not exceed 1% of the shaft diameter. A convenient rate of penetration or uplift for most friction piles is 0·75 mm/min.

96. The hydraulic jack should have a ram movement of at least 15% of the base diameter of the pile plus the estimated elastic shortening or extension of the pile and an allowance for displacement of the reaction system. This allowance may be of the order of 25 mm when piles are used and 100 mm when kentledge is used.

97. It is advisable to specify that a plot of load against penetration or uplift is to be made as the test proceeds so that the approximate point at which ultimate bearing capacity is reached can be identified and the test terminated after an appropriate penetration or uplift has been achieved.

WORK ON COMPLETION OF A PILE TEST

98. Clauses 8.181–8.186 of the model specification are for work to be done on completion of a pile test, but clauses shall be given in the

*See † on page 25.

Particular Specification to define the number of tests to be undertaken and whether test piles, temporary piles and anchorages are to be removed by the Contractor or left for incorporation by him or another contractor in the Permanent Works. If more than one test is specified then the Specification shall cover such matters as the safe-keeping of measuring equipment and kentledge between tests and where the kentledge can be stored on the Site.

MEASUREMENT OF PILING WORK

GENERAL

99.   The procedures in paragraphs 100–156 should be used in conjunction with the appropriate standard method of measurement (CESMM, RICS–NFBTE SMM or DoE SMM). They are of general application whatever the method of measurement, but they are not alternatives to those prescribed under the contract conditions, and should be adapted in accordance with the appropriate standard method of measurement.

100.   These paragraphs explain the intention of the suggested measured items for a bill of quantities given on pages 42–52, and the variations and additions that may be required to meet particular circumstances. The suggested measured items are presented in the order of the types of pile covered in the model specification for piling. Only one item is included for each description. Many contracts will require further items of the same general description to provide for piles of varying section characteristics, lengths and so on.

101.   Materials from which piles are formed and the section characteristics of piles should be stated in item descriptions. Section characteristics are

(*a*) for steel piles, section reference or mass per metre and cross-sectional dimensions

(*b*) for other than steel piles, cross-sectional dimensions or nominal diameters.

Words in parenthesis in the suggested measured items should be substituted by the actual dimensions or details to which they refer.

102.   The relevant standard method of measurement is deemed to prevail except where specifically stated otherwise in the Preamble or in the text of the Bill of Quantities.

103.   The quantity of piling work expressly required and completed is normally measured at completion because of the variability of ground conditions. This presents no difficulty when work is done under the ICE Conditions, the CESMM or the DoE SMM. When carrying out work under a lump sum contract such as the RIBA Conditions and the RIBA–NFBTE SMM, it is customary for piling work to be measured at completion as described above. This should be made clear in the contract documents.

104.   Provision for contingencies should be made by giving Provisional Sums in the Bill of Quantities and not by increasing the quantities beyond those of the work expected to be required (see paragraph 5.17 of the CESMM).

105.   The quantities should be computed net from the Drawings unless directed otherwise in the Contract. No allowance should be made for bulking, shrinkage or waste (see paragraph 5.18 of the CESMM).

PRELIMINARIES

106. The nature of the general and preliminary items to be provided will depend on the items in the Bill of Quantities for the Main Contract. Items to cover general and preliminary works such as the provision of offices and services for the Engineer's Representative, provision of boat and boatmen, are not shown in the suggested list of measured items.

107. Method-Related Charges comprise Time-Related Charges and Fixed Charges as defined in paragraph 7.1 of the CESMM. The distinction between these charges must be made and separate items must be billed for these charges. A contractor tendering in accordance with the CESMM may insert such items for Method-Related Charges as he chooses. Class A of the CESMM provides items for bringing plant to the Site and setting up; the movement of plant around the Site is covered by paragraphs 5.8 and 7.3 of the CESMM. Alternatively the cost of plant may be included in the quantity-related items.

108. Temporary Works, temporary piling and provision of mixing plant may be included as Method-Related Charges.

109. The suggested list of measured items for a bill of quantities indicates a few preliminary items which are Method-Related Charges and which, if not already billed, may be inserted by the tenderer (items A 3 ✽✽ in the CESMM).

INSURANCES

110. A separate item may be provided in the Bill of Quantities to cover insurances generally but this is dependent on the form of the Contract or Sub-contract.

111. Clause 21 of the ICE Conditions states that the Contractor is not liable to insure for the repair or reconstruction of any work constructed with materials and workmanship not in accordance with the requirements of the Contract unless an item has been provided in the Bill of Quantities. An appropriate item, fully described, should be inserted if such insurance is required.

112. No reference to such insurances is made in the RIBA Conditions and if such insurance is required it is necessary to provide an item in the Bill of Quantities and to amend the conditions of the Contract.

SETTING OUT

113. Setting out requirements are given in clause 1.061 of the model specification. A separate item for setting out has not been included in the suggested list of measured items in a bill of quantities but setting out may be measured if desired as part of the item for setting up piling plant at each pile position.

PILING ON LAND

114.　The items in the suggested list have been drafted for use for piling on land.

PILING PLANT

115.　Separate items should be provided in the Bill of Quantities for each required visit. 'Piling plant' means all the plant required to make, install and complete all the piles in the Contract or Sub-contract. A 'rig' means a frame or other piling appliance and the associated plant and labour required to make, install and complete a pile. Items A 3 3 7 . 1 and A 3 3 7 . 2 in the suggested list of measured items may be combined to form one item.

116.　It is desirable to provide additional items when piling is to be installed in separate areas such that the transfer of units of plant from one area to another requires an expenditure of time significantly greater than that required for moving from pile position to pile position in the same area. The separate areas should be defined by the Engineer on the Drawings and in the Specification.

WORK AFFECTED BY WATER

117.　Item descriptions in the Bill of Quantities should distinguish work which is affected by bodies of water (other than groundwater) such as rivers, streams, canals, lakes and tidal water. Item descriptions for work affected by tidal water should also distinguish between work affected at all times and work affected only at some states of the tide. Water surface levels adopted for the purpose of such distinctions should be stated in item descriptions.

WORKING LEVEL

118.　The working level in relation to the Commencing Surface of the ground, river bed or sea bed should be stated in the Bill of Quantities.

119.　Where piling has to be carried out from a fixed structure or staging, whether permanent or temporary, or where floating or temporary staging may be used at the Contractor's discretion, the Bill of Quantities should be drafted accordingly.

120.　'Commencing Surface' means, in relation to an item in a bill of quantities, the surface of the ground before any work covered by the item has been carried out and shall be that adopted in the preparation of the Bill of Quantities as the surface at which boring or driving is expected to begin.

121.　Normally, on land the Commencing Surface will be the ground level at the pile position, and measurement of the pile length to be paid for will start from the Commencing Surface. This includes cases where the piling rig is boring or driving from the normal ground level, or where the rig is standing in the bottom of an excavation from which boring or driving starts. In cases where the rig working level is above the Commencing Surface (e.g. in marine works from a staging or in a cofferdam with the rig at the top of the excavation) the pile length which will be paid for will be measured from the bed level or the bottom of the excavation, in the appropriate depth range measured from working level (see paragraphs 124 and 129).

EXPRESSLY REQUIRED

122. The expression 'expressly required' means shown on the Drawings, described in the Specification or ordered by the Engineer pursuant to the Contract.

DISPLACEMENT OF GROUND

123. Additional length of drive where pile driving causes upward displacement of the ground should not be measured (see clauses 3.117, 5.086, 6.216 and 7.117 of the model specification).

CAST IN PLACE CONCRETE PILES

124. The following separate items are required in the Bill of Quantities for each group of cast in place concrete piles and each cross-section ordered

    (*a*) the number of piles

    (*b*) the total concreted length of the piles

    (*c*) the total depth bored or driven, divided into ranges of depth, there being items for piles which do not exceed 5 m in depth, for those which are longer than 5 m and do not exceed 10 m, and so on in steps of 5 m.

125. The whole depth of boring or driving for each cast in place pile should be included in the item in the range of which the depth occurs.

126. Unconcreted length (empty bore/drive) should be measured at the bore/drive rate.

PILES WITH ENLARGED BASES

127. The diameter of enlarged bases and the diameter of the shaft from which they are enlarged should be stated in item descriptions in the Bill of Quantities.

TREMIE CONCRETE

128. Tremie concrete should be measured in addition to the items listed in paragraph 124. The length measured for this work should be the full length of any pile for which it is used in whole or in part.

PREFORMED CONCRETE, TIMBER AND ISOLATED STEEL PILES

129.   The following separate items are required in the Bill of Quantities for each group of preformed concrete, timber and isolated steel piles and each cross-section ordered

(*a*) the number of piles

(*b*) the total depth driven, undifferentiated as to depth range

(*c*) the total length of piles expressly required to be driven including the length which is required divided into ranges of length, there being items for piles which do not exceed 5 m in length, for those which are longer than 5 m and do not exceed 10 m, and so on in steps of 5 m.

130.   The whole length of each preformed concrete, timber or steel pile should be included in the item in the range of which the length occurs. It should include the length that may be cut off in accordance with the Contract.

MEASUREMENT OF BORED AND DRIVEN DEPTHS

131.   Bored and driven depths should be measured along the axes of piles from the Commencing Surface to the bottom of the shafts of bored piles, to the bottom of the casings of driven cast in place piles and to the bottom of the toes of other driven piles.

LENGTHS OF PILES

132.   The lengths of cast in place piles should be the lengths of in situ concrete including enlarged bases but excluding any cut-off tolerance. The lengths of preformed concrete timber and steel piles should be the lengths expressly required excluding extensions.

RAKING PILES

133.   Separate items, similar to those for vertical piles should be provided in the Bill of Quantities for raking piles. The angle of rake should be stated for piles the angle of rake of which does not exceed 10° and for those where the rake is greater than 10° and does not exceed 20°, and so on in steps of 10°. Measurement should be along the axis of piles. Items should be numbered by means of a suffix on the principle stated in paragraph 4.7 of the CESMM.

PILES FORMED FROM PRECAST UNITS

134.   If precast piles consist of relatively short sections placed one above the other before or during driving, some item descriptions in the Bill of Quantities will have to be omitted or modified and others, adapted to the particular form of construction, may have to be added. Separate items should be provided for each length and cross-section of pile ordered and for other major variations from the Specification.

HANDLING PILES

135.   The handling of piles may be divided into

(a) supply, delivery, unloading and placing in the store yard in sizes and lengths

(b) taking from the store yard, transport and pitching piles.

These details may be included in the items in the Bill of Quantities for the number of piles (see paragraph 129). Where desirable these items may be subdivided.

LENGTHENING

136.   Separate items are required in the Bill of Quantities for the length of pile extensions and for the number of pile extensions. Where piles are extended using cut off material from other piles on the Site items descriptions for the number of pile extensions should so state. The driving of extended piles should be included in the items for driven depth. The preparation of heads to receive pile extensions should not be measured. The length of pile extensions measured should not include lengths formed of material cut from other piles. The lengths measured for timber pile extensions should include lengths occupied by scarfed or other joints.

IN SITU SPLICES OF STEEL PILES

137.   The rates in the Bill of Quantities for splices and the finishes to be applied to them should include material, plant, labour, all preparatory work and delay and, in the case of steel piles, the making good of coatings.

CUTTING OFF AND DISPOSAL

138.   If piling is to be undertaken by a sub-contractor it may be convenient if the cutting off and disposal of surplus lengths of piles is included in the Main Contract. If piling is to be undertaken as a Main Contract cutting off and disposal may be included in the Contract.

OWNERSHIP OF CUT OFF LENGTHS

139.   The ownership of cut off lengths—which is important in the case of steel piles—the approximate place to which they are to be taken and whether they are to be tipped or stacked and subsequently loaded on the Employer's transport or removed from the Site must be stated in the Specification.

DRIVEN CASED PILES

140.   Cast in place piles formed by driving a steel casing or concrete shell in one or more pieces which remain in place after driving and which are filled with concrete should be billed as driven cast in place piles.

PILES CONCRETED IN SITU TO A LEVEL HIGHER THAN CUT-OFF LEVEL

141.   If the concreting of a pile to a level higher than cut-off level and subsequent removal of the additional concrete are specified then the additional concrete should be measured in the rates for the Permanent Works. If the additional concrete is of a different grade from that in the main part of the pile an extra item in the Bill of Quantities should be given.

EXCAVATED SPOIL

142.   The volume of surplus excavated material for disposal should be calculated from the nominal cross-sectional area of the piles and their length measured in accordance with the Contract. The volume of enlarged bases should be added. Disposal of surplus excavated material should be measured only for cast in place piles. 'Disposal' means disposal off the Site unless otherwise stated (see CESMM, note E5).

REINFORCEMENT

143.   Item descriptions in the Bill of Quantities for reinforcement cages should state the number and diameter of the bars of the cages and the material of which they are composed. The rate for reinforcement should include steel in laps, tying wire and additional steel for handling purposes and all spacers of steel or other material. Measurement should include projecting reinforcement to the levels specified.

FILLING TO STEEL PILES

144.   Separate items in the Bill of Quantities should be given for each specified type of filling to steel piles.

CAPS TO STEEL PILES

145.   If the provision of caps to steel piles is specified, e.g. if hollow steel piles are to be filled with an inert liquid, separate items in the Bill of Quantities may be required.

PRELIMINARY PILES

146.   Preliminary piles installed and tested before the start of work on a contract should be measured in a separate bill of quantities or be the subject of a separate contract. Their installation should be measured by the appropriate method of measurement for the type of pile, and testing should be measured in accordance with the suggested measured items for preliminary test loading of piles (page 52).

147.   The Specification should state whether preliminary piles are to remain, to be cut off, to be incorporated in the Permanent Works or

to be withdrawn. Work to be undertaken in connection with the piles should be specified and, if piles are to be withdrawn, ownership and disposal of the piles and making good of the Site should be specified. Separate items in the Bill of Quantities should be included for this work based on similar items for permanent work.

PLANT STANDING DURING TESTS

148. If plant is required to stand during tests of piles payment should be subject to the provisions of clause 36 of the ICE Conditions or clause 6(3) of the RIBA Conditions.

TESTS ON PILES

149. Separate items should be provided in the Bill of Quantities for each specified loading and type of test, e.g. maintained loading, the CRP test and the CRU test. The rates should include all preparatory work, trimming of pile heads and cutting down on completion of the test. If the Engineer requires prices for alternative reaction systems, such as kentledge or anchors, then separate items shall be included for each system. Item coverage should follow paragraphs 146 and 147 as appropriate.

SAMPLES AND IN SITU TESTS

150. Separate items should be provided in the Bill of Quantities for each form of sampling or testing operation, e.g. undisturbed soil samples, the standard penetration test and manufacture of concrete cubes.

DELAYS

151. The model special clauses given in paragraphs 26, 30, 33 and 36 include clauses about payments for delays to plant because of obstruction and for other reasons. Provisional Sum items or references to Dayworks or other rates should be given in the Bill of Quantities to direct how payments for such delays, if approved, shall be made.

BORING THROUGH ROCK

152. If it is expected that rock, weathered rock, boulders and so on will be encountered in bored piling operations a clause should be provided in the Bill of Quantities defining these materials for the purposes of measurement and payment.

PAYMENT BY HOURLY RATES

153. Provision should be made in an appendix to the Bill of Quantities for the Contractor to define his equipment and to insert the normal working hours or shifts on which his tender is based.

OBSERVATIONS AND RECORDS

154. Taking observations, compiling and supplying records in accordance with clause 1.10 of the model specification should be included in the rates.

155. The careful preparation and the maintenance of signed records required in clause 1.10 are important because these records will be used in the measurement of the Works and in the preparation of as-made records of the Works.

SITE INVESTIGATIONS

156. Suggested measured items in a bill of quantities for site investigations are not included in this document because a site investigation project might be covered by a few items in the case of a small site or several hundred items in the case of a major investigation. In the preparation of such a bill, reference should be made to class B of the CESMM.

DRAFT PREAMBLE TO A BILL OF QUANTITIES

Clause A. This Bill of Quantities has been drawn up in accordance with the CESMM/RICS–NFBTE SMM/DoE SMM (delete as appropriate) except that any special methods of measurement used are stated in this Preamble or in the item descriptions for the trades or items affected. All other items are measured net in accordance with the Drawings; no allowance has been made for waste.

*Clause B. The quantities set out in the Bill of Quantities are the estimated quantities of the work but they are not to be taken as the actual and correct quantities of the Works to be executed by the Contractor in fulfilment of his obligations under the Contract. The work actually carried out will be measured at completion and paid for at the rates or prices entered by the Contractor in the Bill of Quantities.

*Clause C. The prices and rates to be inserted in the Bill of Quantities are to be the full inclusive value of the work described under the several items, including all costs and expenses which may be required in and for the construction of the work described together with all general risks, liabilities and obligations set forth or implied in the documents on which the tender is to be based; where special risks, liabilities and obligations cannot be dealt with in this manner, the price thereof shall be stated separately in the item or items which are to be provided for the purpose.

Clause D. General directions and descriptions of work and materials given in the Specification are not necessarily repeated in the Bill of Quantities. Reference shall be made to the Specification for this information. Trade preambles are not necessarily repeated in subsequent bills but shall be held to apply throughout the document.

Clause E. The expression 'expressly required' means shown on the Drawings, described in the Specification or ordered by the Engineer pursuant to the Contract.

Clause F. All work not otherwise described shall be deemed to be performed on land or above MHWS. Work below MLWS is designated as being 'below LWL' and work between MLWS and MHWS as being in tidal range. No warranty as to the accuracy of these levels is given and the Contractor shall allow in his rates for all variations in the tide range. In this connection he should refer to the appropriate tide tables.

Clause G. Payment for standing time for plant will only be made when suspension of work is the Employer's responsibility. Payment for delays due to obstructions where provided for in the Bill of Quantities shall be made at the appropriate rates from the time of reporting the obstruc-

---

*This clause is to be omitted for contracts under the ICE Conditions.

tion to the time when normal working can be resumed, less any time during which the piling plant has been otherwise used. Payment for items for which hourly rates are provided in the Bill of Quantities will be made only for time within the normal working periods as stated by the Contractor in the Bill of Quantities or in his tender.

Clause H. Reference to a unit of piling plant includes all associated plant necessary for the operation of such a unit.

Clause I. The rate for items for number of piles shall include all charges for the setting up of piling plant at each pile, all work on the piles in the store yard, unless otherwise billed, handling of piles from the store yard to the piling plant and pitching the piles and, on completion of work in situ, moving the piling frame to another position.

Clause J. Rates for preparing heads to piles shall include stripping the piles to the specified cut off level and disposing of surplus materials off the Site.

Clause K. The items for lengthening piles shall include stripping the pile heads and/or preparing for lengthening, handling extra lengths of piles and joining them to the piles or casting on an extra length, making good coating, bringing back the piling frame to the lengthened piles, extra driving to move piling and, on completion, moving the piling frame to another position, and shall include for all time lost in so doing.

Clause L. In cases where the rig working level is above the Commencing Surface, the pile length which will be paid for will be measured from the bed level or the bottom of the excavation, in the appropriate depth range measured from working level.

Clause M. The length of pile extensions measured shall not include lengths formed of material arising from cutting off surplus lengths of other piles.

Clause N. The item for driving piles by dollying or with a frame having extended leaders shall be measured as the extra length driven below the Commencing Surface.

Clause O. Excavation around a pile head for work on it shall be measured on Site and the rates shall include for working space and disposal of material off the Site.

Clause P. The rate for the provision of preformed concrete piles shall include all transport, unloading, stacking in the store yard in cross-sectional size and length, all work and labour in the casting yard and plant and materials.

Clause Q. Rates for bored and driven cast in place piles shall include the use of temporary casing.

Clause R. Excavation from within piles shall be measured as the volume of the specified void.

Clause S. Rates for driven cast in place piles shall include any bulb of concrete beyond the nominal diameter and measured length of a pile.

Clause T.   Rates for constructing enlarged bases shall include filling the enlargement with concrete.

Clause U.   The rate for the provision of steel piles shall include all head and toe preparation and the transport, unloading, stacking in the store yard in cross-sectional size and length and, if supplied in lengths shorter than the specified length, the making up of such lengths to the specified lengths.

Clause V.   The rate for the provision of timber piles shall include all transport, unloading, stacking in the store yard in cross-sectional size and length and, unless otherwise billed, all dressing and shaping, checking, drilling and similar work.

Clause W.   The lengths measured for timber pile extensions shall include lengths occupied by scarfed or other joints.

Clause X.   The rate affixed to the items for taking soil samples shall include for all delays to piling operations.

Clause Y.   If a pile longer than that approved is driven payment will not be made for the length of pile which is in excess of the length approved.

SUGGESTED MEASURED ITEMS FOR A BILL OF QUANTITIES

PREFORMED CONCRETE PILES INCLUDING PRESTRESSED CONCRETE PILES

CESMM item number	Item description	Unit	Paragraph†
A 3 3 7.1	Fixed Charge for bringing to Site and erecting piling plant, each visit.	sum	107, 115
A 3 3 7.2	Fixed Charge for dismantling and removing piling plant in item A 3 3 7.1, each visit	sum	107, 115
A 3 3 7.3	Time-Related Charge for keeping on Site plant in item A 3 3 7.1.	sum	107, 115
A 3 3 7.4	Fixed Charge for (number) moves of one unit of piling plant from one defined area to another.	sum	107, 115, 116
P 3–4 * 1	Number of piles (cross-sectional area).	nr	113, 129, 133, 135
P 3–4 * 2.1	Depth driven of (cross-sectional area) pile.	m	129, 131, 133
P 3–4 * 2.2	Depth driven of (cross-sectional area) pile by dollying or with frame having extended leaders.	m	131, 133
P 3–4 * 2.3	Depth driven of lengthened pile (cross-sectional area).	m	131, 136
P 3–4 * 3–8	Length of pile (cross-sectional area, length range).	m	129
Q 2 1 *	Pre-bore for pile (cross-sectional area) and dispose of soil arising.	m	152
Q 2 4 *	Lengthen pile (cross-sectional area).	nr	136
Q 2 5 *	Length of pile extension (cross-sectional area).	m	136
Q 2 6 *	Cut off surplus length (cross-sectional area).	nr	138, 139
Q 6 2 0	Standing by of driving rig when expressly required, per rig.	h	151, 153

†These paragraphs are to be excluded from contract documents.

CESMM item number			Item description	Unit	Paragraph†
Q	6	7	0	Pre-boring through artificial obstructions for (cross-sectional dimensions) pile when expressly required, per rig.	h	152
E	3	4	*	Excavating around pile head when expressly required and disposal of spoil.	m³	142

†These paragraphs are to be excluded from contract documents.

BORED CAST IN PLACE PILES

CESMM item number				Item description	Unit	Paragraph†
A	3	3	8.1	Fixed Charge for bringing to Site and erecting piling plant, each visit.	sum	107, 115
A	3	3	8.2	Fixed Charge for dismantling and removing piling plant in item A 3 3 8.1, each visit.	sum	107, 115
A	3	3	8.3	Time-Related Charge for keeping on Site plant in item A 3 3 8.1	sum	107, 115
A	3	3	8.4	Fixed Charge for (number) moves of one unit of piling plant from one defined area to another.	sum	107, 115, 116
P	1	*	1	Number of (diameter) piles.	nr	113, 124, 133
P	1	*	2	Length of piles concreted (diameter).	m	124–126, 132, 133
P	1	*	3–8	Depth bored of (diameter) pile (depth range).	m	124–126, 131, 133, 152
Q	1	3	*	Backfilling empty bore (diameter) with (stated material).	m	
Q	1	4	*	Permanent casing to pile (diameter and thickness).	m	
Q	1	5	*	Placing concrete by tremie in (diameter) pile.	m	128
Q	1	6	*	Enlarged bases (diameter) to (shaft diameter) piles.	nr	127, 142
Q	1	7	*	Preparing heads to (diameter) piles.	nr	
Q	1	8	0	Reinforcement to pile (numbers, diameter and length bars per pile).	t	143
Q	1	9	0	Temporary casing to pile (diameter) left in when expressly required.	m	
Q	6	6	0	Standing by of boring rig when expressly required, per rig.	h	151, 153
Q	6	7	0	Boring through artificial obstructions when expressly required, per rig.	h	152

†These paragraphs are to be excluded from contract documents.

CESMM item number	Item description	Unit	Paragraph†
Q 7 0 0	Surplus excavated material for disposal.	m³	142
B 4 2 0	Undisturbed samples (diameter) of cohesive material taken from pile boreholes.	nr	150
B 6 2 5	Laboratory tests: set of undrained triaxial compression tests (number in set) on (diameter) samples including all handling and transport.	nr	150, 152

†These paragraphs are to be excluded from contract documents.

CESMM item number				Item description	Unit	Paragraph†
A	3	3	7.1	Fixed Charge for bringing to Site and erecting piling plant, each visit.	sum	107, 115
A	3	3	7.2	Fixed Charge for dismantling and removing piling plant in item A 3 3 7.1, each visit.	sum	107, 115
A	3	3	7.3	Time-Related Charge for keeping on Site plant in item A 3 3 7.1.	sum	107, 115
A	3	3	7.4	Fixed Charge for (number) moves of one unit of piling plant from one defined area to another.	sum	107, 115, 116
P	2	*	1	Number of (diameter) piles.	nr	113, 124–126, 129, 133, 135, 140
P	2	*	2	Length of piles concreted (diameter).	m	124–126, 132, 133, 140
P	2	*	3–8	Depth driven of (diameter) pile (depth range).	m	124–126, 131–133, 140
Q	1	1	*	Pre-boring.	m	152
Q	1	3	*	Backfilling empty drive (diameter) with (stated material).	m	
Q	1	4	*	Permanent casing to pile (diameter and thickness).	m	140
Q	1	5	*	Placing concrete by tremie in (diameter) pile.	m	128
Q	1	7	*	Preparing heads to (diameter) piles.	nr	138
Q	1	8	0	Reinforcement to pile (number, diameter and length bars per pile).	m	143
Q	1	9	0	Temporary casing to pile (diameter) left in when expressly required.	m	
Q	6	1	0	Standing by of driving rig when expressly required, per rig.	h	151, 153

†These paragraphs are to be excluded from contract documents.

CESMM item number				Item description	Unit	Paragraph†
Q	6	7	0	Boring through artificial obstructions when expressly required, per rig.	h	152
Q	7	0	0	Surplus excavated material for disposal.	m³	142

†These paragraphs are to be excluded from contract documents.

ISOLATED STEEL BEARING PILES

CESMM item number	Item description	Unit	Paragraph†
A 3 3 7.1	Fixed Charge for bringing to Site and erecting piling plant, each visit.	sum	107, 115
A 3 3 7.2	Fixed Charge for dismantling and removing piling plant in item A 3 3 7.1, each visit.	sum	107, 115
A 3 3 7.3	Time-Related Charge for keeping on site plant in item A 3 3 7.1.	sum	107, 115
A 3 3 7.4	Fixed Charge for (number) moves of one unit of piling plant from one defined area to another.	sum	107, 115, 116
P 7 * 1	Number of piles (cross-sectional dimensions, mass per metre, length with shoe).	nr	113, 129, 133, 135
P 7 * 2.1	Depth driven of (cross-sectional dimensions, mass per metre) pile.	m	129, 133
P 7 * 2.2	Depth driven of (cross-sectional area) pile by dollying or with frame having extended leaders.	m	129, 133, 136
P 7 * 2.3	Depth driven of lengthened pile (cross-sectional dimensions, mass per metre).	m	131, 133
P 7 * 3–8	Length of pile (cross-sectional dimensions, mass per metre, length range).	m	132
Q 4 1 *	Pre-bore for pile (cross-sectional dimensions) and dispose of spoil arising.	m	152
Q 4 4 *	Lengthen pile (cross-sectional dimensions, mass per metre).	nr	136, 137
Q 4 5 *	Length of pile extension (cross-sectional dimensions, mass per metre).	m	136
Q 4 6 *	Cut off surplus length after driving (cross-sectional dimensions and mass per metre) and stack cut off at (place) for re-use.	nr	138, 139
Q 4 3.1 *	Place concrete by tremie (grade) in pile (cross-sectional dimensions).	m	144

†These paragraphs are to be excluded from contract documents.

48

CESMM item number	Item description	Unit	Paragraph†
Q 4 3.2 *	Place concrete by tremie (grade) in pile (cross-sectional dimensions).	m	128
Q 4 9.1 *	Fill pile (cross-sectional dimensions) with (liquid).	m	144, 145
Q 4 9.2 *	Provide and weld (class) steel cap (dimensions and thickness) to (cross-sectional dimensions) pile.	nr	145
Q 6 4 0	Standing by of driving rig when expressly required, per rig.	h	153
Q 6 7 0	Bore through artificial obstructions for pile (cross-sectional dimensions) when expressly required, per rig.	h	152
A 3 1 9.1	Fixed Charge for bringing to Site and erecting covered structure under which coating of piles is undertaken.	sum	109
A 3 1 9.2	Fixed Charge for dismantling and removing from Site covered structure in A 3 1 9.1	sum	109
A 3 1 9.3	Fixed Charge for keeping on Site covered structure in A 3 1 9.1	sum	109
M 7 1 0	Blast cleaning (cross-sectional dimensions) pile.	m²	
M 7 7 0	Paint (number) coats of primer and (number) coats of (type) paint the whole having minimum total thickness (z mm).	m²	
E 3 4 1–5	Excavating around pile head when expressly required and diposal of spoil.	m³	142

†These paragraphs are to be excluded from contract documents.

CESMM item number				Item description	Unit	Paragraph†
A	3	3	7.1	Fixed Charge for bringing to Site and erecting piling plant, each visit.	sum	107, 115
A	3	3	7.2	Fixed Charge for dismantling and removing piling plant in item A 3 3 7.1, each visit.	sum	107, 115
A	3	3	7.3	Time-Related Charge for keeping on Site plant in item A 3 3 7.1.	sum	107, 115
A	3	3	7.4	Fixed Charge for (number) moves of one unit of piling plant from one defined area to another.	sum	107, 115, 116
P	6	1–5	1.1	Number of piles (cross-sectional area).	nr	113, 129, 133, 135
P	6	1–5	2.1	Depth driven of (cross-sectional area) pile.	m	129, 131, 133
P	6	1–5	2.2	Depth driven of (cross-sectional area) pile by dollying or with frame having extended leaders.	m	131, 133
P	6	1–5	2.3	Depth driven of lengthened pile (cross-sectional area).	m	131, 136
P	6	1–5	3–8	Length of pile (cross-sectional area length range).	m	129
Q	3	1	1–5	Pre-bore for pile (cross-sectional area) and dispose of spoil arising.	m	152
Q	3	4	1–5	Lengthen pile (cross-sectional area) including re-use of ring or new ring.	nr	136
Q	3	5	1–5	Length of pile extension (cross-sectional area).	m	136
Q	3	6	1–5	Cutting off surplus length (cross-sectional area) pile and stacking cut-offs at (place) for re-use.	nr	138, 139
Q	3	9	1–5	Preparing toe of pile (cross-sectional area) and fixing cast steel shoe (weight including fittings).	nr	

†These paragraphs are to be excluded from contract documents.

CESMM item number				Item description	Unit	Paragraph†
Q	3	7	1–5	Preparing head of pile (cross-sectional area) and fixing (material) ring (thickness, depth and fittings).	nr	
Q	6	3	0	Standing by of driving rig when expressly required, per rig.	h	153
Q	6	7	0	Pre-boring through artificial obstructions for (cross-sectional area) pile, per rig.	h	152
E	3	4	✳	Excavating around pile head when expressly required and disposal of spoil.	m³	142

†These paragraphs are to be excluded from contract documents.

PRELIMINARY TEST LOADING OF PILES
(paragraphs 146 and 147)

CESMM item number				Item description	Unit	Paragraph†
A	3	3	7.1	Fixed Charge for bringing to Site all piling equipment, stagings, jacks, beams, straps to connect same to piles to form jack reaction, supports, test gauges and other measuring instruments and all other equipment necessary for lateral loading, tension test loading and compression test loading as specified but excluding kentledge, and removing from Site when no longer required (tests measured separately).	sum	107, 115
	or					
A	3	3	8.1			
A	3	3	7.2	Time-Related Charge for keeping on Site piling test equipment provided under item A 3 3 7.1	sum	107, 115
	or					
A	3	3	8.2			
A	3	3	7.3	Fixed Charge for bringing to site and removing kentledge (t).	sum	107, 115
	or					
A	3	3	8.3			
A	3	3	7.4	Time-Related Charge for keeping on Site kentledge (t).	sum	107, 115
	or					
A	3	3	8.4			
Q	6	✱	0	Standing by or moving of rig when expressly required, per rig.	h	153
Q	6	7	0	Bore through obstructions when expressly required, per rig.	h	152
Q	8	1–2	1–8.1	Erect and dismantle testing equipment and carry out maintained loading to a maximum test load (t) and unloading tests.	nr	113, 149
Q	8	1–2	1–8.2	Erect and dismantle testing equipment and carry out constant rate of penetration tests to a maximum test load (t).	nr	113, 149
Q	8	1–2	1–8.3	Erect and dismantle equipment and carry out lateral loading to a maximum test load (t).	nr	113, 149
Q	8	1–2	1–8.4	Erect and dismantle equipment and carry out tension tests to a maximum test load (t).	nr	113

†These paragraphs are to be excluded from contract documents.

52

MODEL SPECIFICATION FOR PILING

SECTION 1. GENERAL REQUIREMENTS FOR PILING WORK

1.01. STANDARDS AND DEFINITIONS

1.011. British Standard specifications

All materials and workmanship shall be in accordance with the appropriate current British Standards, including those listed in clause 1.12, except that where the requirements of British Standards are in conflict with this specification, the latter shall take precedence.

1.012. Codes of practice

All work shall be carried out generally in accordance with the principles of relevant codes of practice, including those listed in clause 1.12.

1.013. Definitions

In this specification the terms 'approved', 'approval' and 'required' mean 'approved by the Engineer', 'approval of the Engineer' and 'required by the Engineer' respectively.

1.02. GENERAL CONTRACT REQUIREMENTS

1.021. Particular Specification

The following matters, where appropriate, are described in the Particular Specification

(*a*) nature of the Works
(*b*) classes of loads on piles
(*c*) contract drawings
(*d*) other works proceeding at the same time
(*e*) working area
(*f*) order of the Works
(*g*) datum
(*h*) offices for the Engineer's Representative
(*i*) details of soil investigation reports

1.022. Contractor to work to other contractors' drawings

The Contractor, where so directed by the Engineer, shall be required to work to other contractors' drawings whenever drawings for works not included in the Contract are related to particular details of the Works.

1.023. Temporary fencing

The Contractor shall provide all temporary fencing required by him in connection with the execution of the Contract.

1.024. Advertisements

The Contractor shall allow no advertisement to be placed on any hoarding, scaffolding or fencing erected in connection with the Contract without the permission of the Engineer.

1.025. Life-saving appliances

The Contractor shall provide and maintain on the Site sufficient, proper and efficient life-saving appliances to the approval of the Engineer. The appliances must be conspicuous and available for use at all times.

1.026. Progress report

The Contractor shall submit to the Engineer on the first day of each week, or at such longer periods as the Engineer may from time to time direct, a progress report showing the current rate of progress and progress during the previous period on all important items of each section of the Works.

1.027. Latrine and washing accommodation

The Contractor shall provide and maintain to the satisfaction of the Engineer and of the appropriate statutory authority all adequate, efficient and sanitary latrine and washing accommodation required for the use of his staff and work people employed on the Site, with proper attendance, drainage and water supply. He shall, when it is no longer required, remove such latrine and washing accommodation and generally restore the Site to a clean and sanitary condition.

1.028. Telephone facilities

The Contractor shall make his own arrangements with the appropriate authorities for the provision of telephone facilities to and on the Site.

1.029. Flammable stores

All petroleum, explosives and flammable materials shall be stored in fireproof buildings and such precautions taken with regard to siting and fire risks as the Engineer may direct. The Contractor shall make all arrangements with the licensing authority for the necessary licence.

1.03. MATERIALS

1.031. Sources of supply

The sources of supply of materials shall not be changed without prior approval.

1.032. Rejected materials

Rejected materials are to be removed promptly from the Site.

1.04. SAFETY

1.041. Standards

Safety precautions throughout the piling operations shall comply with the Health and Safety at Work Act 1974 or any subsequent re-enactment thereof, and with CP 2004 and CP 2011.

1.042. Diving

Diving operations shall be carried out in accordance with *The diving operations special regulations 1960** and any subsequent amendments or additions thereto.

Before any diving is undertaken the Contractor shall supply the Engineer with two copies of the code of signals to be employed, and shall have a copy of that code prominently displayed on the craft or structure from which any diving operation takes place.

1.05. SOIL CONDITIONS

1.051. Soil investigation reports

Factual information and reports on site investigations for the Works shall be made available to tenderers. However, no responsibility is accepted by the Employer for any opinions or conclusions given in the reports.

Before the start of work the Contractor shall be given a copy of the foregoing information and any further information which may have been obtained.

1.052. Unexpected ground conditions

The Contractor shall report immediately to the Engineer any circumstance which indicates that in the Contractor's opinion the ground conditions differ from those expected by him from his interpretation of the site investigation reports.

1.06. TOLERANCES

1.061. Setting out

Setting out shall be carried out from the main grid lines of the proposed structure. Immediately before installation of the pile, the pile position shall be marked with suitable identifiable pins or markers.

1.062. Position

For a pile cut off at or above ground level the maximum permitted deviation of the pile centre from the centre point shown on the setting out drawing shall be 75 mm in any direction. An additional tolerance for a pile head cut off below ground level will be permitted in accordance with clauses 1.063 and 1.064.

1.063. Verticality

The maximum permitted deviation of the finished pile from the vertical is 1 in 75.

1.064. Rake

The piling rig shall be set and maintained to attain the required rake. The maximum permitted deviation of the finished pile from the specified rake is 1 in 25.

---

* HMSO, London, Statutory Instrument 688.

1.065. Forcible corrections

Forcible corrections to concrete piles shall not be made. Forcible corrections may be made to other piles only if approved.

1.07. PILING METHOD

The Contractor shall supply for approval all relevant details of the method of piling and the plant he proposes to use. Any alternative method to that specified shall be in accordance with the Specification.

1.08. PERFORMANCE SPECIFICATION

Where piling is required to meet a performance specification the Contractor shall, when tendering, supply full details of the type of pile offered, the standards of control he intends to use, how the calculation and checking of the load bearing capacity and settlement of the piles will be carried out, and the tests he proposes to undertake on the Site.

1.09. PILING PROGRAMME

The Contractor shall inform the Engineer each day of the programme of piling for the following day and shall give adequate notice of his intention to work outside normal hours and at weekends.

1.10. RECORDS

The Contractor shall keep records as indicated by an asterisk in Table 1 of the installation of each pile and shall submit two signed copies of these records to the Engineer not later than noon of the next working day after the pile was installed. The signed records will form a record of the work.

Any unexpected driving or boring conditions reported in accordance with clauses 1.052, 3.115, 5.084, 6.214 and 7.114 shall be briefly noted in the records.

1.11. NUISANCE AND DAMAGE
1.111. Noise and disturbance

The Contractor shall carry out the work in such a manner and at such times as to minimize noise and disturbance.

1.112. Damage to adjacent structures

If during the execution of the work damage is, or is likely to be, caused to mains, services or adjacent structures, the Contractor shall submit to the Engineer his proposals for repair or avoidance of such damage.

1.113. Damage to piles

The Contractor shall ensure that damage does not occur to completed piles.

The Contractor shall submit to the Engineer his proposed sequence and timing for driving or boring piles having regard to the avoidance of damage to adjacent piles.

Table 1

Data	Driven pre-cast steel and timber piles	Bored cast in place piles	Driven cast in place piles
(a) Contract	*	*	*
(b) Pile reference number (location)	*	*	*
(c) Pile type	*	*	*
(d) Nominal cross-sectional dimensions or diameter	*	*	*
(e) Nominal diameter of underream	—	*	—
(f) Length of preformed pile	*	—	—
(g) Standing groundwater level	—	*	—
(h) Date and time of driving, redriving or boring	*	*	*
(i) Date of concreting	—	*	*
(j) Ground level at commencement of installation of pile	*	*	*
(k) Working level	*	*	*
(l) Depth from working level to pile toe	*	*	*
(m) Toe level	*	*	*
(n) Depth from working level to pile head level	—	*	*
(o) Length of temporary casing	—	*	*
(p) Length of permanent casing	—	*	*
(q) Type, weight, drop and mechanical condition of hammer and equivalent information for other equipment	*	—	*
(r) Number and type of packings used and type and condition of dolly used during driving the pile	*	—	*
(s) Set of pile or pile tube in mm per 10 blows or number of blows per 25 mm of penetration	*	—	*
(t) If required, the sets taken at intervals during the last 3 m of driving	*	—	*
(u) If required, temporary compression of ground and pile from time of a marked increase in driving resistance until pile reaches its final level	*	—	*
(v) Soil samples taken and in situ tests carried out	—	*	—
(w) Length and details of reinforcement	—	*	*
(x) Concrete mix	—	*	*
(y) Volume of concrete supplied to pile where this can be measured in practice	—	*	*
(z) All information regarding obstructions delays and other interruptions to the sequence of work	*	*	*

1.114. *Temporary support*

The Contractor shall ensure that piles are temporarily braced or stayed immediately after driving to prevent loosening of the piles in the ground and to ensure that no damage resulting from oscillation, vibration or movement of any free-standing pile length can occur.

1.12. *RELEVANT STANDARDS AND CODES OF PRACTICE*

Aggregates, cement and concrete

British Standards

12	Portland cement (ordinary and rapid hardening)
812	Methods for sampling and testing of mineral aggregates, sands and fillers
882, 1201	Aggregates from natural sources for concrete (including granolithic)
1047	Air-cooled blast furnace slag coarse aggregates for concrete
1305	Batch type concrete mixers
1881 (part 4)	Methods of testing concrete for strength
1926	Ready-mixed concrete
3148	Tests for water for making concrete
3963	Methods for testing the mixing performance of concrete mixers
4027	Sulphate-resisting Portland cement
4251	Truck type concrete mixers

British standard codes of practice

110	The structural use of concrete
114	Structural use of reinforced concrete in buildings
115	The structural use of prestressed concrete in buildings
116	The structural use of precast concrete

Iron and steel

British Standards

4	Structural steel sections
449	The use of structural steel in building
1452	Grey iron castings
1775	Steel tubes for mechanical, structural and general engineering purposes
2691	Steel wire for prestressed concrete
3100	Steel castings for general engineering purposes
3617	Seven wire steel strand for prestressed concrete
4232	Surface finish of blast-cleaned steel for painting
4360	Weldable structural steels
4449	Hot rolled steel bars for reinforcement of concrete
4461	Cold worked steel bars for reinforcement of concrete

4466	Bending dimensions and scheduling of bars for the reinforcement of concrete
4486	Cold worked high tensile alloy steel bars for prestressed concrete

British Standard code of practice

CP 2008	Protection of iron and steel structures from corrosion

Swedish Standard

SIS 05 59 00	Surface preparation for painting steel surfaces

Welding

British Standards

499	Welding terms and symbols
638	Arc welding plant, equipment and accessories
639	Covered electrodes for the manual metal-arc welding of mild steel and medium-tensile steel
693	General requirements for oxy-acetylene welding of mild steel
709	Methods of testing fusion welded joints and weld metal in steel
938	General requirements for the metal-arc welding of structural steel tubes to BS 1775
1719	Classification, coding and marking of covered electrodes for metal-arc welding
1856	General requirements for the metal-arc welding of mild steel
2642	General requirements for the arc welding of carbon manganese steels
2645	Tests for use in the approval of welders
2937	General requirements for seam welding in mild steel
4165	Electrode wires and fluxes for the submerged arc welding of carbon steel and medium-tensile steel

American standards

ASME	Boiler and pressure vessel Code Section 8
API 5L	Line pipe
API 1104	Welding pipelines and related facilities

Timber

British Standards

144	Coal tar creosote for preservation of timber
565	Glossary of terms relating to timber and woodwork
881, 589	Nomenclature of commercial timbers, including sources of supply

60

913	Wood preservation by means of pressure creosoting
1282	Classification of wood preservatives and their method of application
4072	Wood preservation by means of water-borne copper/chrome/arsenic compositions

British standard codes of practice

| 98 | Preservative treatment for constructional timber |
| 112 | The structural use of timber |

Soil mechanics and foundation engineering

British Standard

| 1377 | Methods of testing soils for civil engineering purposes |

British standard codes of practice

2001	Site investigation
2004	Foundations
2011	Safety precautions in the construction of large diameter boreholes for piling and other purposes

Building Research Establishment Digest

| 174 | Concrete in sulphate-bearing soils and ground waters |

Geological Society's Engineering Group Working Party Report

The logging of rock cores for engineering purposes

Oil Companies Materials Association, London

| DFCP 4 | Drilling fluid materials—bentonite |

Safety

British Standards

| 229 | Flameproof enclosure of electrical apparatus |
| 2011 | Safety precautions in the construction of large diameter boreholes for piling and other purposes |

SECTION 2. GENERAL REQUIREMENTS FOR CONCRETE PILES

2.01. *GENERAL*

All materials shall be in accordance with section 1 of this specification, the Particular Specification and this section, except where there may be conflict of requirements, in which case those in the Particular Specification and this section shall take precedence.

2.02. *CEMENT*

2.021. *Type of cement*

Cement shall be ordinary Portland, rapid-hardening Portland, sulphate-resisting Portland, super-sulphated or Portland blast-furnace cement unless otherwise specified. When forwarding his piling method and programme to the Engineer, the Contractor shall submit for approval the type of cement, other than ordinary Portland cement, he proposes to use. High alumina cement shall not be used.

2.022. *Storage of cement*

All cement shall be stored in separate containers according to type in substantially built waterproof stores or silos.

2.03. *AGGREGATE*

2.031. *Types of aggregate*

Aggregates shall consist of naturally occurring material unless otherwise specified or ordered. The Contractor shall inform the Engineer of the source of supply of the aggregates before the commencement of work and, at the request of the Engineer, provide evidence regarding their properties and consistency.

The content of chloride salt in the aggregate used in steel reinforced concrete work, expressed as the equivalent anhydrous calcium chloride percentage by weight of the cement to be used in the concrete, shall be less than 1%.

The content of chloride salt in the aggregate used in pre-tensioned prestressed concrete work, expressed as the equivalent anhydrous calcium chloride percentage by weight of the cement to be used in the concrete, shall be less than 0·1%. The limit of 0·1% shall also apply to the main concrete of post-tensioned prestressed concrete work unless there is an impermeable and durable barrier, in addition to any grout, between the main concrete and the tendons.

The use of marine aggregates will not normally be approved for use other than with ordinary or sulphate-resisting Portland cement.

Should the use of calcium chloride as an additive be approved, the amount used shall be such that the limits of 1% for steel reinforced concrete and 0·1% for pre-tensioned prestressed concrete work and post-tensioned prestressed concrete work are not exceeded.

2.032. Storage of aggregates

All aggregates brought to the Site shall be free and kept free from deleterious matter. Aggregates of different types and sizes shall be stored separately in different hoppers or different stockpiles.

2.04. WATER

2.041. Availability

If water for the Works is not available from a public supply, approval shall be obtained regarding the source of water.

2.042. Tests

When required by the Engineer, the Contractor shall arrange for tests of the water for the Works to be carried out in accordance with BS 3148 before and during the progress of the work.

2.05. ADMIXTURES

Admixtures may be used if approved and shall be used when required.

2.06. CONCRETE MIXES

2.061. Grade designation

Grades of concrete shall be denoted by the characteristic 28 day test cube strength in newtons per square millimetre.

2.062. Mix

Concrete mixes shall be in accordance with clause 2.063 (designed mix) or clause 2.064 (prescribed mix) and of grades 20, 25, 30, 40 or 50 of CP 110 or other grades approved appropriate to the work.

The Contractor shall submit the slump factor he proposes for approval before work commences. Neither trial mixes nor strength tests are required for prescribed mixes.

The concrete shall have sufficient workability to enable it to be placed and compacted by the methods used in forming the piles specified in sections 3–5.

2.063. Designed mix

When a designed mix is specified the Contractor shall be responsible for selecting the mix proportions to achieve the required strength and workability, but the Engineer will be responsible for specifying the minimum cement content and any other properties required to ensure durability.

Designed mixes shall be in accordance with grades 20, 25, 30, 40 or 50 of CP 110 or other grades approved appropriate to the work. Complete information on the mix and sources of aggregate for each grade of concrete and the water/cement ratio and the proposed degree of workability shall be approved before work commences.

63

2.064. Prescribed mix

When a prescribed mix is specified the Engineer will specify the mix proportions and the Contractor shall undertake to provide a properly mixed concrete containing the constituents in the specified proportions.

Prescribed mixes shall be in accordance with Table 2 or Table 3. Table 2 is applicable to concrete which can be compacted in position by the use of internal vibrators or by other approved methods. Table 3, based on the use of gravel aggregates, is applicable to concrete for use in case in place piling, where high workability is essential and where the use of internal vibrators is precluded.

Table 2. Prescribed mixes for piling

Concrete grade	Nominal maximum size of aggregate, mm	40		20	
	Workability	Medium	High	Medium	High
	Slump limits, mm	50–100	100–150	25–75	75–125
20	Cement, kg/m³	300	320	320	350
	Total aggregate, kg/m³	1850	1750	1800	1750
	Sand/total aggregate: zone 1	35%	40%	40%	45%
	zone 2	30%	35%	35%	40%
	zone 3	30%	30%	30%	35%
25	Cement, kg/m³	340	360	360	390
	Total aggregate, kg/m³	1800	1750	1750	1700
	Sand/total aggregate: zone 1	35%	40%	40%	45%
	zone 2	30%	35%	35%	40%
	zone 3	30%	30%	30%	35%
30	Cement, kg/m³	370	390	400	430
	Total aggregate, kg/m³	1750	1700	1700	1650
	Sand/total aggregate: zone 1	35%	40%	40%	45%
	zone 2	30%	35%	35%	40%
	zone 3	30%	30%	30%	35%

Notes.

1. If the specific gravity of either the coarse or the fine aggregate differs significantly from 2·6, the weight of each type of aggregate shall be adjusted in proportion to the specific gravity of the materials.

2. If a crushed stone sand or a crushed gravel sand is used instead of natural sand the weight of the coarse aggregate shall be reduced by at least 90 kg without alteration to the weight of sand or cement.

3. The cement shall be ordinary Portland cement or sulphate-resisting Portland cement. Where other cements listed in clause 2.021 are used, special consideration shall be given to the design of the mix.

4. The weights of cement and dry aggregates are those which will produce approximately one cubic metre of compacted concrete.

2.065. Sulphate attack

For concrete in piles exposed to sulphate attack the requirements of Building Research Establishment Digest 174 shall be observed.

2.066. Minimum cement content

The cement content in any mix shall be not less than 300 kg/m³. Where concrete is to be placed under water or drilling mud by tremie, or where the pile will be exposed to sea water, the cement content shall be not less than 400 kg/m³.

2.067. Maximum cement content

The cement content in any mix shall not exceed 550 kg/m³.

Table 3. *Prescribed mixes for cast in place piling using only gravel aggregates of nominal maximum size 20 mm*

Concrete grade	Piling mix workability	A	B	C
	Slump limits, mm	75–125	100–175	150 or over
20	Cement, kg/m³	320	350	400
	Total aggregate, kg/m³	1840	1790	1740
	Sand/total aggregate: zone 1	40%	—	—
	zone 2	35%	40%	40%
	zone 3	32%	35%	35%
25	Cement, kg/m³	360	400	460
	Total aggregate, kg/m³	1790	1740	1680
	Sand/total aggregate: zone 1	40%	—	—
	zone 2	35%	40%	40%
	zone 3	32%	35%	35%
30	Cement, kg/m³	410	460	520
	Total aggregate, kg/m³	1750	1680	1600
	Sand/total aggregate: zone 1	40%	—	—
	zone 2	35%	40%	40%
	zone 3	32%	35%	35%

Notes.
1. This table is based on the use of gravel aggregates, and is not applicable to crushed rock aggregates.
2. If the specific gravity of either the coarse or the fine aggregate differs significantly from 2·6, the weight of each type of aggregate shall be adjusted in proportion to the specific gravity of the materials.
3. The cement shall be ordinary Portland cement or sulphate-resisting Portland cement. Where other cements listed in clause 2.021 are used, special consideration shall be given to the design of the mix.
4. The weights of cement and dry aggregates are those which will produce approximately one cubic metre of compacted concrete.

2.07. TRIAL MIXES

2.071. General

When designed mixes are specified trial mixes shall be prepared for each grade of concrete in accordance with BS 1881, unless there are existing data showing that the proposed mix proportions and manufacture will produce a concrete of the strength and quality required having adequate workability for compaction by the method to be used in placing.

2.072. Preliminary trial mixes

When required in accordance with clause 2.071 the Contractor shall, before the commencement of concreting, have preliminary trial mixes prepared, preferably under full-scale production conditions or, if this is not possible, in an approved laboratory using a sufficient number of samples to be representative of the aggregates and cement to be used. Unless otherwise approved for each grade of concrete a set of six cubes shall be made from each of three batches in accordance with clause 6.5.3 of CP 110. From each set of six cubes three shall be tested at an age of 7 days and three at 28 days.

This procedure shall be followed when accelerated testing is proposed for works cubes, but an additional three cubes from each batch shall be made, cured and tested in accordance with the accelerated regime.

2.073. Trial mixes during the work

Where a trial mix is required after commencement of the work the procedure in clause 2.072 shall be adopted for full-scale production conditions as approved. The strength requirement shall be as in clause 2.075.

2.074. Workability

The workability of each batch of the trial mixes shall be determined by the slump test as described in BS 1881 or by an alternative approved method.

2.075. Standard of acceptance

Unless otherwise approved, the trial mix proportions will be approved if the criteria in clause 6.5.3, paragraph 2, of CP 110 are met.

2.076. Variations in approved mix

When a mix has been approved, no variations shall be made in the proportions, the original source of the cement and aggregates or their type, size or grading zone without the consent of the Engineer. Further tests may be required.

2.08. TESTING WORKS CONCRETE

2.081. Sampling

Concrete for the piles shall be sampled in accordance with BS 1881.

2.082. Workability

The workability of concrete shall be determined by the slump test as described in BS 1881 or by an alternative approved method.

2.083. Works cube tests

For each grade of concrete four cubes shall be made from a single batch when required for each 50 m³ of concrete or part thereof in each day's work. The cubes shall be made, cured and tested in accordance with BS 1881 or as otherwise approved. One shall be tested at an age of 7 days and the other three at 28 days. Alternatively cubes may be tested in accordance with an approved accelerated testing regime. The Contractor shall submit certified copies of the results of all tests to the Engineer.

2.084. Standard of acceptance

The standard of acceptance of the works cubes shall be in accordance with clause 6.8.2.2 of CP 110 or as otherwise approved. Where the scale and duration of the Contract is small the Contractor shall submit for approval an alternative standard of acceptance before the commencement of work.

2.085. Records of tests

The Contractor shall keep a detailed record of the results of all tests on concrete and concrete materials. Each test shall be clearly identified with the piles to which it relates.

2.09. BATCHING CONCRETE

2.091. General

Unless otherwise specified the requirements in clauses 2.092, 2.093 and 2.094 shall be met.

2.092. Accuracy of weighing and measuring equipment

The weighing and water-dispensing mechanisms shall be maintained at all times to within the limits of accuracy described in BS 1305.

2.093. Tolerance in weights

The weights of the quantities of each size of aggregate and of cement shall be within a tolerance of 2% of the respective weights per batch after due allowance has been made for the presence of free water in the aggregates which shall be determined by the Contractor by an approved method.

67

2.094. Moisture content of aggregates

The moisture content of aggregates shall be measured immediately before mixing and as frequently thereafter as is necessary to maintain consistency of mix.

2.10. MIXING CONCRETE
2.101. Type of mixer

The mixer shall be of the batch type, unless otherwise approved, and shall have either been manufactured in accordance with BS 1305 or shown by tests in accordance with BS 3963 to have mixing performance within the limits of table 6 of BS 1305.

2.102. Tolerance of mixer blades

The mixing blades of pan mixers shall be maintained within the tolerances specified by the manufacturers of the mixers, and the blades shall be replaced when it is no longer possible to maintain the tolerances by adjustment.

2.103. Cleaning of mixers

Mixers which have been out of use for more than 30 minutes shall be thoroughly cleaned before another batch of concrete is mixed. Unless otherwise specified by the Engineer, the first batch of concrete through a mixer shall contain the normal batch quantity of cement and sand, but only two thirds of the normal quantity of coarse aggregate. Mixing plant shall be thoroughly cleaned between the mixing of different types of cement.

2.104. Minimum temperature

The temperature of fresh concrete shall not be allowed to fall below 5°C. No frozen material or materials containing ice shall be used.

2.11. TRANSPORTING CONCRETE
2.111. Method of transporting

The method of transporting concrete shall be submitted for approval. Concrete shall be transported in uncontaminated watertight containers in such a manner that loss of material and segregation are prevented.

2.112. Pumping concrete

The use of pumped concrete and the methods employed in its use shall be subject to approval.

2.12. READY-MIXED CONCRETE
2.121. Conditions for use

Subject to approval the Contractor may use ready-mixed concrete in accordance with BS 1926. Approval shall be obtained for each proposed use of ready-mixed concrete in different sections of the

Works and for each different mix, which shall comply with this specification.

2.122. Mixing plant
Unless otherwise agreed by the Engineer, truck mixer units and their mixing and discharge performance shall comply with the requirements of BS 4251.

2.13. STEEL REINFORCEMENT
2.131. Condition
Steel reinforcement shall be stored in clean conditions. It shall be clean, and free from loose rust and loose mill scale at the time of fixing in position and subsequent concreting.

2.132. Grade
The grade of steel shall be as specified.

2.133. Bending of reinforcement
No reinforcement shall be bent at a temperature lower than 5°C without prior approval. Before bending, reinforcement may be warmed to a temperature not exceeding 100°C. If reinforcement already cast into concrete has to be bent, the internal radius of the bend shall be not less than twice the diameter of bars of mild steel or three times the diameter of bars of high-yield steel.

2.134. Placing of reinforcement
All intersecting bars shall be tied together with approved wire unless otherwise permitted by the Engineer.

Reinforcement in the form of a cage shall be assembled with additional support, such as spreader forks and lacings, necessary to form a rigid cage. Hoops, links or helical reinforcement shall fit closely around the main longitudinal bars and be bound to them by approved wire, the ends of which shall be turned into the interior of the pile or pour. Reinforcement shall be placed and maintained in position.

The cover to all reinforcement shall be not less than 40 mm thick.

Spacer blocks shall be approved and shall be as small as possible consistent with their purpose. They shall be made with 10 mm maximum aggregate size and shall have a strength not less than that of the concrete in the pile.

2.135. Welding of reinforcement
Welded joints and welding procedures shall be carried out in accordance with BS 693 or BS 1856.

SECTION 3. PRECAST NORMAL REINFORCED AND PRESTRESSED PILES

3.01. *GENERAL*

All materials and work shall be in accordance with sections 1 and 2 of this specification, the Particular Specification and this section, except where there may be conflict of requirements, in which case those in the Particular Specification and this section shall take precedence.

3.02. *PARTICULAR SPECIFICATION*

The following matters are, where appropriate, described in the Particular Specification

(*a*) types of cement
(*b*) types of aggregate
(*c*) grades of concrete
(*d*) designed or prescribed mixes
(*e*) specified strength
(*f*) grades and types of reinforcement
(*g*) types of prestressing tendons
(*h*) grout
(*i*) marking of piles
(*j*) driving resistance
(*k*) designed loads
(*l*) penetration
(*m*) disposal of cut-off lengths.

Materials

3.03. *PILE SHOES*

Cast iron pile shoes shall be made from chill hardened iron as used for making grey iron castings to BS 1452, grade 10. The chilled iron point shall be free from major blow holes and other surface defects.

Steel pile shoes shall be manufactured from steel to BS 4360, grade 43AI.

Cast steel piles shoes shall be of steel to BS 3100, grade A.

Straps or other fastenings to cast pile shoes shall be of steel to BS 4360, grade 43A, and shall be cast into the point to form an integral part of the shoe.

Workmanship

3.04. *TOLERANCES IN PILE DIMENSIONS*

The cross-sectional dimensions of the pile shall be not less than those specified and shall not exceed them by more than 6 mm.

Any face of a pile shall not deviate by more than 6 mm from a straight edge 3 m long laid on the face, and the centroid of any cross-section of the pile shall not deviate by more than 12 mm from the straight line connecting the centroids of the end faces of the pile.

3.05. REINFORCEMENT

The main longitudinal reinforcing bars in piles not exceeding 12 m in length shall be in one continuous length unless otherwise specified. In piles exceeding 12 m long, joints will be permitted in main longitudinal bars at 12 m nominal intervals. Joints in adjacent bars shall be staggered at least 1 m apart along the length of the pile. Joints in reinforcement shall be such that the full strength of the bar is effective across the joint.

Lap or splice joints shall be provided with sufficient link bars to resist eccentric forces.

Welded joints shall be made in accordance with BS 693 or BS 1856.

3.06. FORMWORK

The Contractor shall state in his tender if piles are to be cast using adjacent piles as forms. When the sides of adjacent piles are used as formwork, an approved method shall be used to prevent adhesion between concrete surfaces.

The head of each pile shall be square to the longitudinal axis. The corners of the head and the corners of the pile shaft for a distance of 300 mm from the head shall be chamfered 25 mm × 25 mm.

The point of the pile or shoe shall lie on the longitudinal axis of the pile.

Holes for toggle bolts shall be at right angles to the faces of the piles, and shall be lined with steel tubes or other approved material. Holes for handling and pitching shall be provided and shall be lined with steel tubes; alternatively, approved inserts may be cast in.

Formwork shall be sufficiently robust in construction to withstand the disturbances due to placing and compacting of concrete, and shall be such that there shall be no loss of material from wet concrete. The method of forming hollow cores shall be such that a continuous core is formed.

All formwork shall be clean immediately before concrete is placed.

Formwork shall be struck not less than 12 hours after the completion of the compaction of the concrete.

Formwork shall be removed without damage to the concrete.

3.07. CONCRETE
3.071. Compacting concrete

Unless otherwise agreed by the Engineer, concrete shall be compacted with the assistance of vibrators. Internal vibrators shall be capable of producing not less than 150 Hz and external vibrators not less than 50 Hz. Internal vibrators shall operate not closer than 75 mm to shuttering.

Every care shall be taken during compaction to prevent segregation of the constituents of concrete and the displacement of reinforcement.

3.072. *Protecting and curing concrete*

Immediately after compaction concrete shall be adequately protected from the harmful effects of the weather, including wind, rain, rapid temperature changes and frost. It shall also be protected from drying out by an approved method of curing.

The period of curing at an ambient temperature of 10°C shall be not less than that shown in Table 4, increased or decreased by the Engineer to achieve the necessary maturity if the temperature is respectively less than or greater than 10°C.

Table 4

Type of cement	Wet curing time after completion of placing concrete
Ordinary Portland	4 days
Sulphate-resisting Portland	4 days
Portland blast-furnace	4 days
Super-sulphated	4 days
Rapid-hardening Portland	3 days

When accelerated curing is used the curing procedure shall be approved. Four hours must elapse from the completion of placing concrete before the temperature is raised. The rise in temperature within any period of 30 minutes shall not exceed 10 deg C and the maximum temperature attained shall not exceed 70°C. The rate of subsequent cooling shall not exceed the rate of heating.

3.08. PRESTRESSING

3.081. *General*

Tensioning shall be carried out only when the Engineer is present, unless otherwise approved. In cases where piles are manufactured off the Site the Contractor shall ensure that the Engineer is given adequate notice and every facility for inspecting the manufacturing process.

Prestressing operations shall be carried out only under the direction of an experienced and competent supervisor. All personnel operating the stressing equipment shall have been trained in its use.

72

3.082. Concrete strength

The Contractor shall cast sufficient cubes to be able to demonstrate by testing two cubes at a time, at approved intervals between pairs of cubes, that the specified transfer strength of the concrete has been reached. Unless otherwise permitted in the Specification, concrete shall not be stressed until two test cubes attain the specified transfer strength.

The calculated extensions and total forces, including allowance for losses, shall be agreed with the Engineer before stressing is commenced.

Stressing of tendons and transfer of prestress shall be carried out at a gradual and steady rate. The force in the tendons shall be obtained from readings on a load cell or pressure gauge incorporated in the equipment. The extension of the tendons under the agreed total forces shall be within 5% of the agreed calculated extension.

3.083. Records

The Contractor shall keep detailed records of times of tensioning, measured extensions, pressure gauge readings or load cell readings and the amount of pull-in at each anchorage. Copies of these records shall be supplied to the Engineer within such reasonable time of completion of each tensioning operation as may be required, and in any case not later than noon on the following working day.

3.084. Grout

Unless otherwise directed or agreed by the Engineer the grout shall

(*a*) consist only of ordinary Portland cement, water and approved additives

(*b*) have a water/cement ratio as low as possible consistent with the necessary workability, but the water/cement ratio shall not exceed 0·45 unless an approved mix containing an expanding agent is used

(*c*) not be subject to bleeding in excess of 2% after 3 hours or 4% maximum when measured at 18°C in a covered glass cylinder approximately 100 mm in diameter with a height of grout of approximately 100 mm, and the water shall be reabsorbed after 24 hours.

Admixtures containing chlorides or nitrates shall not be used.

3.085. Post-tensioned piles

Ducts and vents in post-tensioned piles shall be grouted after the transfer of prestress.

3.086. Grouting procedure

Grout shall be mixed for a minimum of 2 minutes and until a uniform consistency is obtained.

73

Ducts shall not be grouted when the air temperature in the shade is lower than 3°C.

Before grouting is started all ducts shall be thoroughly cleaned by means of compressed air. In the case of vertical piles this shall be done from the bottom upwards.

Grout shall be injected near the lowest point in the duct in one continuous operation and allowed to flow from the outlet until the consistency is equivalent to that of the grout being injected. Vents in ducts shall be provided in accordance with clause 6.12.3 of CP 110.

3.087. Records

The Contractors shall keep records of grouting including the date, the proportions of the grout and any admixtures used, the pressure, details of interruption and topping required. Copies of these records shall be supplied to the Engineer within such reasonable time after completion of each grouting operation as may be required and in any case not later than noon on the following working day.

3.09. MARKING OF PILES

After a pile has been cast, the date of casting, reference number, length and, where appropriate, the prestressing force shall be clearly inscribed on the top surface of the pile and also clearly and indelibly marked on the head of the pile. In addition, each pile shall be marked at intervals of 250 mm along the top 3 m of its length before being driven.

3.10. HANDLING AND STORAGE OF PILES

The method and sequence of lifting, handling, transporting and storing piles shall be such that the piles are not damaged. Only the designed lifting and support points shall be used. During transport and storage piles shall be stored on adequate supports located under the lifting points of the piles.

Concrete shall at no time be subjected to loading, including its own weight, which will induce a compressive stress in it exceeding 0.33 of its strength at the time of loading or of the specified strength, whichever is the lesser. For this purpose the assessment of the strength of the concrete and of the stresses produced by the loads shall be subject to the agreement of the Engineer.

All piles within a stack shall be in groups of the same length. Packings of uniform thickness shall be provided between piles at the lifting points.

3.11. DRIVING PILES

3.111. Strength of piles

Piles shall not be driven until the concrete has achieved the specified characteristic strength.

3.112. Leaders and trestles

At all stages during driving and until incorporation in the superstructure the pile shall be adequately supported and restrained by means of leaders, trestles, temporary supports or other guide arrangements to maintain position and alignment and to prevent buckling. These arrangements shall be such that damage to the pile does not occur.

3.113. Performance of driving equipment

The Contractor shall satisfy the Engineer regarding the suitability, efficiency and energy of the driving equipment. Unless otherwise approved, drop hammers shall not be used from floating craft.

3.114. Length of piles

The length of pile to be driven in any position shall be approved.

3.115. Driving procedure and redrive checks

Each pile shall be driven continuously until the specified or approved set and/or depth has been reached, except that the Engineer may permit the suspension of driving if he is satisfied that the rate of penetration prior to the cessation of driving will be substantially re-established on its resumption or if he is satisfied that the suspension of driving is beyond the control of the Contractor. A follower (long dolly) shall not be used unless approved, in which case the Engineer will require the set to be revised to take into account the reduction in the effectiveness of the hammer blow.

The Contractor shall inform the Engineer without delay if an unexpected change in driving characteristics is noted. A detailed record of the driving resistance over the full length of the nearest available pile shall be taken if required.

At the start of work and in a new area or section sets shall be taken at intervals during the last 3 m of the driving to establish the behaviour of the piles.

The Contractor shall give adequate notice and provide all facilities to enable the Engineer to check driving resistances. A set shall be taken only in the presence of the Engineer unless otherwise approved.

Redrive checks, if required, shall be carried out to an approved procedure.

3.116. Final set

The final set of each pile shall be recorded either as the penetration in millimetres per 10 blows or as the number of blows required to produce a penetration of 25 mm.

When a final set is being measured, the following requirements shall be met.

(a) The exposed part of the pile shall be in good condition without damage or distortion.

(*b*) The dolly and packing, if any, shall be in sound condition.

(*c*) The hammer blow shall be in line with the pile axis and the impact surfaces shall be flat and at right angles to the pile and hammer axis.

(*d*) The hammer shall be in good condition and operating correctly.

(*e*) The temporary compression of the pile shall be recorded if required.

3.117. *Driving sequence and risen piles*

Piles shall be driven in an approved sequence to minimize the detrimental effects of heave and lateral displacement of the ground.

When required, levels and measurements shall be taken to determine the movement of the ground or any pile resulting from the driving process.

When a pile has risen as a result of adjacent piles being driven, the Contractor shall submit to the Engineer his proposals for correcting this and the avoidance of it in subsequent work.

3.118. *Preboring*

If preboring is specified the pile shall be pitched into a hole prebored to the depth shown on the Drawings.

3.119. *Jetting*

Jetting shall be carried out only when the Contractor's detailed proposals have been approved and not over the last 3 m of penetration.

3.12. *REPAIR AND LENGTHENING OF PILES*

3.121. *Repair of damaged pile heads*

When repairing the head of a pile, the head shall be cut off square at sound concrete, and all loose particles shall be removed by wire brushing, followed by washing with water. If the pile is to be subjected to further driving, the head shall be replaced by concrete of an approved grade.

If the driving of a pile has been accepted but sound concrete of the pile is below the cut-off level, the pile shall be made good to the cut-off level with concrete of a grade not inferior to that of the concrete of the pile.

3.122. *Lengthening of normal reinforced piles*

When lengthening a normal reinforced pile, the head shall be cut off square to sound concrete, and all loose particles shall be removed by wire brushing, followed by washing with water.

Joints in reinforcement shall be such that the full strength of the bar is effective across the joint.

Welded joints shall be made in accordance with BS 693 or BS 1856 and the main longitudinal reinforcing bars in the head of the pile shall be exposed for at least 300 mm below the weld.

For lap or splice joints sufficient link bars shall be provided to resist eccentric forces.

If the pile is to be subjected to further driving the additional length shall be of an approved grade. Other methods of lengthening shall be subject to approval.

3.123. *Lengthening of prestressed piles*

If mild steel bars have been designed and incorporated in the head of a prestressed pile during manufacture for use as bonding bars, the pile shall be lengthened, if required, in the manner described in clause 3.122.

Any provision for lengthening prestressed piles incorporated at the time of manufacture shall be as designed or approved.

If no provision for lengthening prestressed piles was incorporated at the time of manufacture, any method for lengthening shall be such that joints are capable of taking safely the stresses during driving and under load.

3.124. *Driving repaired or lengthened piles*

Repaired or lengthened piles shall not be driven until the added concrete has reached the specified characteristic strength of the concrete of the pile.

3.13. *CUTTING OFF PILE HEADS*

When the driving of a pile has been approved, and if not otherwise specified, the concrete of the head of the pile shall be cut off to the level specified or shown on the Drawings. The length of splice reinforcing bars projecting above this level shall be as specified on the Drawings.

Care shall be taken to avoid shattering or otherwise damaging the rest of the pile. Any cracked or defective concrete shall be cut away and made good with new concrete properly bonded to the old.

SECTION 4. BORED CAST IN PLACE PILES

4.01. GENERAL

All materials and work shall be in accordance with sections 1 and 2 of this specification, the Particular Specification and this section, except where there may be conflict of requirements, in which case those in the Particular Specification and in this section shall take precedence.

4.02. PARTICULAR SPECIFICATION

The following matters are, where appropriate, described in the Particular Specification

- (*a*) samples and testing
- (*b*) types of cement
- (*c*) types of aggregate
- (*d*) grades of concrete
- (*e*) designed or prescribed mixes
- (*f*) grades and types of reinforcement
- (*g*) permanent casing
- (*h*) bentonite
- (*i*) designed loads.

4.03. SOIL SAMPLES

Soil, rock or groundwater samples shall be taken, or soil tests carried out in situ, while the pile is being bored. The samples shall be taken to an approved laboratory for testing as specified.

The taking of samples and all subsequent handling and transporting shall be carried out in accordance with CP 2001.

Materials

4.04. PERMANENT CASINGS

Permanent casings shall be as specified.

4.05. DRILLING FLUID
4.051. Supply

Bentonite, as supplied to the Site and prior to mixing, shall be in accordance with specification DFCP 4 of the Oil Companies Materials Association (see clause 1.12).

A certificate shall be obtained by the Contractor from the manufacturer of the bentonite powder, showing the properties of each consignment delivered to the Site. This certificate shall be made available to the Engineer on request. The properties to be given by the manufacturer are the apparent viscosity range (in centipoises) and the gel strength range (in newtons per square metre) for solids in water.

4.052. Mixing

Bentonite shall be mixed thoroughly with clean fresh water to make a suspension which will maintain the stability of the pile excavation for the period necessary to place concrete and complete construction. The temperature of the water used in mixing the bentonite suspension, and of the suspension when supplied to the borehole, shall be not lower than 5°C.

Where saline or chemically contaminated groundwater occurs, special precautions shall be taken to modify the bentonite suspension or prehydrate the bentonite in fresh water so as to render it suitable in all respects for the construction of piles.

4.053. Tests

The frequency of testing drilling fluid and the method and procedure of sampling shall be proposed by the Contractor prior to the commencement of the work. The frequency may subsequently be varied as required, depending on the consistency of the results obtained.

Control tests shall be carried out on the bentonite suspension, using suitable apparatus. The density of freshly mixed bentonite suspension shall be measured daily as a check on the quality of the suspension being formed. The measuring device shall be calibrated to read to within 0·005 g/ml. Tests to determine density, viscosity, shear strength and pH value shall be applied to bentonite supplied to the pile boring. For average soil conditions the results shall generally be within the ranges stated in Table 5. The tests shall be carried out until a consistent

Table 5

Property to be measured	Range of results at 20°C	Test method
Density	Less than 1·10 g/ml	Mud density balance
Viscosity	30–90 s or less than 20 cP	Marsh cone method Fann viscometer*
Shear strength (10 minute gel strength)	1·4–10 N/m² or 4–40 N/m²	Shearometer Fann viscometer
pH	9·5–12	pH indicator paper strips or electrical pH meter

* Where the Fann viscometer is specified, the fluid sample should be screened by a number 52 sieve (300 μm) prior to testing.

working pattern has been established, account being taken of the mixing process, any blending of freshly mixed bentonite suspension and previously used bentonite suspension and any process which may be used to remove impurities from previously used bentonite suspension. When the results show consistent behaviour, the tests for shear strength and pH value may be discontinued, and tests to determine density and viscosity shall be carried out as agreed with the Engineer. In the event of a change in the established working pattern, tests for shear strength and pH value shall be reintroduced for a period if required.

Workmanship

4.06. SETTING OUT

The Contractor shall check the casing position for each pile during and immediately after placing the casing, and agree it with the Engineer.

4.07. DIAMETER OF PILES

The diameter of a pile shall be not less than the specified diameter.

4.08. BORING

4.081. Boring near recently cast piles

Piles shall not be bored so close to other piles which have recently been cast and which contain workable or unset concrete that a flow of concrete could be induced from or damage caused to any of the piles.

4.082. Temporary casings

Temporary casing of approved quality or an approved alternative method shall be used to maintain the stability of pile excavation which might otherwise collapse.

Temporary casings shall be free from significant distortion. They shall be of uniform cross-section throughout each continuous length. During concreting they shall be free from internal projections and en-crusted concrete which might prevent the proper formation of piles.

4.083. Stability of pile excavation using drilling fluid

Where the use of drilling fluid is approved for maintaining the stability of a boring, the level of the fluid in the excavation shall be maintained so that the fluid pressure always exceeds the pressures exerted by the soils and external groundwater, and an adequate temporary casing shall be used in conjunction with the method to ensure stability of the strata near ground level until concrete has been placed. The fluid level shall be maintained at a level not less than 1 m above the level of the external groundwater.

In the event of a rapid loss of bentonite suspension from the pile excavation, the excavation shall be backfilled without delay and the instructions of the Engineer shall be obtained before excavation at that location is resumed.

4.084. Spillage and disposal

All reasonable steps shall be taken to prevent the spillage of bentonite suspension on the Site in areas outside the immediate vicinity of boring. Discarded bentonite shall be removed from the Site without delay. Any disposal of bentonite shall comply with the regulations of the local controlling authority.

4.085. Pumping from boreholes

Pumping from a borehole shall not be permitted unless a casing has been placed into a stable stratum which prevents the flow of water from other strata in significant quantities into the boring, or unless it can be shown that pumping will not have a detrimental effect on the surrounding soil or property.

4.086. Continuity of construction

A pile constructed in a stable cohesive soil without the use of temporary casing or other form of support shall be bored and concreted without prolonged delay and in any case soon enough to ensure that the soil characteristics are not significantly impaired.

4.087. Enlarged pile bases

An enlarged base mechanically formed shall be not smaller than the dimensions specified and shall be concentric with the pile shaft to within a tolerance of 10% of the shaft diameter. The sloping surface of the frustrum forming the enlargement shall make an angle to the horizontal of not less than 55°.

4.088. Cleanliness of pile bases

On completion of boring, loose, disturbed or remoulded soil shall be removed from the base of the pile.

4.089. Inspection

Each pile boring shall be inspected prior to the placing of concrete in it. This inspection shall be carried out from the ground surface in the case of dry bores of diameter of less than 750 mm. Where the diameter of a dry boring is 750 mm or greater, equipment shall be provided to enable the Contractor and the Engineer to descend into the boring for the purpose of inspection. Any method of descent and the equipment used shall comply with CP 2011.

4.09. Reinforcement

Joints in longitudinal steel bars will be permitted unless otherwise specified. Joints in reinforcement shall be such that the full strength of the bar is effective across the joint and shall be made so that there is no relative displacement of the reinforcement during the construction of the pile.

4.10. PLACING CONCRETE

4.101. General

The method of placing and the workability of the concrete shall be such that a continuous monolithic concrete shaft of the full cross-section is formed.

The concrete shall be placed without such interruption as would allow the previously placed batch to have hardened. The method of placing shall be approved.

The Contractor shall take all precautions in the design of the mix and placing of the concrete to avoid arching of the concrete in a casing. No spoil, liquid or other foreign matter shall be allowed to contaminate the concrete.

4.102. Workability of concrete

Slump measured at the time of discharge into the pile boring shall be in accordance with the standards shown in Table 6.

Concrete shall be of the workability approved under clause 2.062

Table 6

Piling mix workability	Slump		Typical conditions of use
	Minimum, mm	Maximum, mm	
A	75	125	Placed into water-free unlined bore. Widely spaced reinforcement leaving ample room for free movement between bars
B	100	175	Where reinforcement is not spaced widely enough to give free movement between bars. Where casting level of concrete is within the casing. Where pile diameter is less than 600 mm
C	150		Where concrete is to be placed by tremie under water or drilling fluid

when in its final position and after all constructional procedures in forming the pile have been completed.

4.103. Compaction

Internal vibrators shall not be used to compact concrete unless the Contractor is satisfied that they will not cause segregation or arching of the concrete and unless the method of use has been approved.

4.104. Placing concrete in dry borings

Approved measures shall be taken to avoid segregation and bleeding and to ensure that the concrete at the bottom of the pile is not deficient in grout.

4.105. Placing concrete under water or drilling fluid

Concrete to be placed under water or drilling fluid shall be placed by tremie unless otherwise approved and shall not be discharged freely into the water or drilling fluid.

Before placing concrete, measures shall be taken to ensure that there is no accumulation of silt or other material at the base of the boring and the Contractor shall ensure that heavily contaminated bentonite suspension, which could impair the free flow of concrete from the pipe of the tremie, has not accumulated in the bottom of the hole.

A sample of the bentonite suspension shall be taken from the base of the boring using an approved sampling device. If the specific gravity of the suspension exceeds 1·25 the placing of concrete shall not proceed. In this event the Contractor shall modify or replace the bentonite as approved to meet the Specification.

The concrete shall be a rich coherent mix of high workability in accordance with clause 2.066 and Table 6, mix C.

The concrete shall be placed in such a manner that segregation does not occur.

During and after concreting care shall be taken to avoid damage to the concrete from pumping and dewatering operations.

The hopper and pipe of the tremie shall be clean and watertight throughout. The pipe shall extend to the base of the boring and a sliding plug or barrier shall be placed in the pipe to prevent direct contact between the first charge of concrete in the pipe of the tremie and the water or drilling fluid. The pipe shall at all times penetrate the concrete which has previously been placed and shall not be withdrawn from the concrete until completion of concreting. At all times a sufficient quantity of concrete shall be maintained within the pipe to ensure that the pressure from it exceeds that from the water or drilling fluid. The internal diameter of the pipe of the tremie shall be not less than 150 mm for concrete made with 20 mm aggregate and not less than 200 mm for concrete made with 40 mm aggregate. It shall be so designed that external projections are minimized, allowing the tremie to pass through reinforcing cages without causing damage. The internal face of the pipe of the tremie shall be free from projections.

4.11. EXTRACTION OF CASING
4.111. Workability of concrete
Temporary casings shall be extracted while the concrete within them remains sufficiently workable to ensure that the concrete is not lifted.

4.112. Concrete level
When the casing is being extracted a sufficient quantity of concrete shall be maintained within it to ensure that pressure from external water, drilling fluid or soil is exceeded and that the pile is neither reduced in section nor contaminated.

No concrete shall be placed in the boring once the bottom of the casing has been lifted above the top of the concrete; it shall be placed continuously as the casing is extracted until the desired head of concrete is obtained.

Adequate precautions shall be taken in all cases where excess heads of water or drilling fluid could be caused as the casing is withdrawn because of the displacement of water or fluid by the concrete as it flows into its final position against the walls of the shaft. Where two or more discontinuous lengths of casing (double casing) are used in the construction the proposed method of working shall be approved.

4.113. Vibrating extractors
The use of vibrating casing extractors will be permitted subject to clauses 1.111 and 1.112.

4.114. Water levels
In the event of the groundwater level being higher than the required pile head casting level shown on the Drawings, the Contractor shall submit his proposals for approval prior to placing concrete. The pile head shall not be left below the groundwater level unless approved precautions are taken.

4.12. TEMPORARY BACKFILLING ABOVE PILE CASTING LEVEL
After each pile has been cast any empty bore remaining shall be protected and shall be carefully backfilled as soon as possible with approved materials.

SECTION 5. DRIVEN CAST IN PLACE PILES

5.01. GENERAL

All materials and work shall be in accordance with sections 1 and 2 of this specification, the Particular Specification and this section, except where there may be conflict of requirements, in which case those in the Particular Specification and in this section shall take precedence.

This section shall apply to piles for which a permanent casing of steel or concrete is driven and which may be reinforced prior to being filled with concrete or for which a temporary casing is driven, reinforcement placed in the temporary casing and the hole formed by the temporary casing filled with concrete during extraction of the casing.

5.02. PARTICULAR SPECIFICATION

The following matters are, where appropriate, described in the Particular Specification

- (*a*) types of cement
- (*b*) types of aggregate
- (*c*) grades of concrete
- (*d*) designed or prescribed mixes
- (*e*) specified strength
- (*f*) grades and types of reinforcement
- (*g*) types and quality of permanent casing
- (*h*) types and quality of pile shoes
- (*i*) driving resistance
- (*j*) penetration
- (*k*) designed load.

Materials

5.03. PERMANENT CASINGS

Permanent casings shall be as specified. Where a permanent casing is to be made from a series of short sections it shall be designed and placed so as to produce a continuous waterfree shaft. The dimensions and quality of the casing shall be adequate to withstand the stresses caused by handling and driving without damage or distortion.

5.04. PILE SHOES

Pile shoes shall be manufactured from durable material capable of withstanding the stresses caused by driving without damage, and shall be designed to give a watertight joint when used with any permanent casing.

85

5.05. DIAMETER OF PILES

The diameter of a pile shall be not less than the specified diameter.

5.06. TEMPORARY CASINGS

Temporary casings shall be free from significant distortion. They shall be of uniform cross-section throughout each continuous length. During concreting they shall be free from internal projections and encrusted concrete which might prevent the proper formation of piles.

5.07. ENLARGED PILE BASES

Where the Contractor wishes to form a pile with an enlarged base, details of the proposed method of forming the base and the materials to be used shall be submitted at the time of tendering.

5.08. DRIVING PILES

5.081. *Piling near recently cast piles*

Casings shall not be driven or piles formed so close to other piles which have recently been cast and which contain workable or unset concrete that a flow of concrete could be induced from or damage caused to any of the piles.

5.082. *Performance of driving equipment*

The Contractor shall satisfy the Engineer regarding the suitability, efficiency and energy of the driving equipment. Unless otherwise approved, drop hammers shall not be used from floating craft.

5.083. *Length of piles*

The length of pile to be driven in any position shall be approved.

5.084. *Driving procedure and redrive checks*

Each pile shall be driven continuously until the specified or approved set and/or depth has been reached, except that the Engineer may permit the suspension of driving if he is satisfied that the rate of penetration prior to the cessation of driving will be substantially reestablished on its resumption or if he is satisfied that the suspension of driving is beyond the control of the Contractor. A follower (long dolly) shall not be used unless approved, in which case the Engineer will require the set to be revised to take into account the reduction in the effectiveness of the hammer blow.

The Contractor shall inform the Engineer without delay if an unexpected change in driving characteristics is noted. A detailed record of the driving resistance over the full length of the nearest available pile shall be taken if required.

At the start of work and in a new area or section sets shall be taken at intervals during the last 3 m of the driving to establish the behaviour of the piles.

The Contractor shall give adequate notice and provide all facilities to enable the Engineer to check driving resistances. A set shall be taken only in the presence of the Engineer unless otherwise approved.

Redrive checks, if required, shall be carried out to an approved procedure.

5.085. *Final set*

The final set of each pile shall be recorded either as the penetration in millimetres per 10 blows or as the number of blows required to produce a penetration of 25 mm.

When a final set is being measured, the following requirements shall be met.

(*a*) The exposed part of the pile shall be in good condition without damage or distortion.

(*b*) The dolly and packing, if any, shall be in sound condition.

(*c*) The hammer blow shall be in line with the pile axis and the impact surfaces shall be flat and at right angles to the pile and hammer axis.

(*d*) The hammer shall be in good condition and operating correctly.

(*e*) The temporary compression of the pile shall be recorded if required.

5.086. *Driving sequence and risen piles*

Piles shall be driven in an approved sequence to minimize the detrimental effects of heave and lateral displacement of the ground.

When required, levels and measurements shall be taken to determine the movement of the ground or any pile resulting from the driving process.

When a pile has risen as a result of adjacent piles being driven, the Contractor shall submit to the Engineer his proposals for correcting this and the avoidance of it in subsequent work.

5.087. *Preboring*

If preboring is specified the pile shall be pitched into a hole prebored to the depth shown on the Drawings.

5.088. *Jetting*

Jetting shall be carried out only when the Contractor's detailed proposals have been approved, and not over the last 3 m of penetration.

5.089. *Internal drop hammer*

Where a casing for a pile without an enlarged base is to be driven by an internal drop hammer a plug consisting of concrete grade 20 with a water/cement ratio not exceeding 0·25 shall be placed in the pile.

This plug shall have a compacted height of not less than $2\frac{1}{2}$ times the diameter of the pile. Fresh concrete shall be added to ensure that this height of driving plug is maintained in the casing throughout the period of driving and in any event a plug of fresh concrete shall be added after $1\frac{1}{2}$ hours of normal driving, after $\frac{3}{4}$ hour of hard driving or, should the pile driving be interrupted, fresh concrete shall be added after $\frac{1}{2}$ hour of the cessation of driving.

5.09. REPAIR OF DAMAGED PILE HEADS

When repairing the head of a pile, the head shall be cut off square at sound concrete, and all loose particles shall be removed by wire brushing, followed by washing with water.

If the driving of a pile has been accepted but sound concrete of the pile is below the cut-off level, the pile shall be made good to the cut-off level with concrete of a grade not inferior to that of the concrete of the pile.

5.10. LENGTHENING OF CAST IN PLACE PILES

When lengthening a cast in place pile, the head shall be cut off square to sound concrete, and all loose particles shall be removed by wire brushing, followed by washing with water.

Joints in reinforcement shall be such that the full strength of the bar is effective across the joint.

Welded joints shall be made in accordance with BS 693 or BS 1856 and the main longitudinal reinforcing bars in the head of the pile shall be exposed for at least 300 mm below the weld.

For lap or splice joints sufficient link bars shall be provided to resist eccentric forces.

If the pile is to be subjected to further driving the additional length shall be of an approved grade. Other methods of lengthening shall be subject to approval.

5.11. INSPECTION

Prior to placing concrete in a pile casing the Contractor shall check in an approved manner that any permanent casing is undamaged, and that the casing is free from water or other foreign matter. In the event of water or foreign matter having entered the pile casing either the casing shall be withdrawn, repaired if necessary and redriven or other action shall be taken as may be approved to continue the construction of the pile.

5.12. REINFORCEMENT

Joints in longitudinal steel bars will be permitted unless otherwise specified. Joints in reinforcement shall be such that the full strength of the bar is effective across the joint and shall be made so that there is no relative displacement of the reinforcement during the construction of the pile.

5.13. PLACING CONCRETE

5.131. General

The method of placing and the workability of the concrete shall be such that a continuous monolithic concrete shaft of the full cross-section is formed.

The concrete shall be placed without such interruption as would allow the previously placed batch to have hardened. The method of placing shall be approved.

The Contractor shall take all precautions in the design of the mix and placing of the concrete to avoid arching of the concrete in a casing. No spoil, liquid or other foreign matter shall be allowed to contaminate the concrete.

5.132. Workability of concrete

Slump measured at the time of discharge into the pile boring shall be in accordance with the standards shown in Table 7, except that those mixes shall not apply to piling systems which use semi-dry concrete and employ special means for its compaction.

The concrete shall be of the workability approved under clause 2.062 when in its final position and after all constructional procedures in forming the pile have been completed.

Table 7

Piling mix workability	Slump		Typical conditions of use
	Minimum, mm	Maximum, mm	
A	75	125	Placed into water-free unlined hole. Widely spaced reinforcement leaving ample room for free movement between bars
B	100	175	Where reinforcement is not spaced widely enough to give free movement between bars. Where casting level of concrete is within the casing. Where pile diameter is less then 600 mm
C	150		Where concrete is to be placed by tremie under water or drilling fluid

5.133. Compaction

Internal vibrators shall not be used to compact concrete unless the Contractor is satisfied that they will not cause segregation or arching of the concrete and unless the method of use has been approved.

5.134. Placing concrete in dry borings

Approved measures shall be taken to avoid segregation and bleeding and to ensure that the concrete at the bottom of the pile is not deficient in grout.

5.135. Placing concrete under water

Concrete to be placed under water shall be placed by tremie unless otherwise approved and shall not be discharged freely into the water.

Before placing concrete, measures shall be taken to ensure that there is no accumulation of silt or other material at the base of the boring.

The concrete shall be a rich coherent mix of high workability in accordance with clauses 2.066 and Table 7, mix C.

The concrete shall be placed in such a manner that segregation does not occur.

During and after concreting care shall be taken to avoid damage to the concrete from pumping and dewatering operations.

The hopper and pipe of the tremie shall be clean and watertight throughout. The pipe shall extend to the base of the pile and a sliding plug or barrier shall be placed in the pipe to prevent direct contact between the first charge of concrete in the pipe of the tremie and the water. The pipe shall at all times penetrate the concrete which has previously been placed and shall not be withdrawn from the concrete until completion of concreting. At all times a sufficient quantity of concrete shall be maintained within the pipe to ensure that the pressure from it exceeds that from the water. The internal diameter of the tremie pipe shall be not less than 150 mm for concrete made with 20 mm aggregate and not less than 200 mm for concrete made with 40 mm aggregate. It shall be so designed that external projections are minimized, allowing the tremie to pass through reinforcing cages without causing damage. The internal face of the pipe of the tremie shall be free from projections.

5.14. EXTRACTION OF CASING

5.141. Workability of concrete

Temporary casings shall be extracted while the concrete within them remains sufficiently workable to ensure that the concrete is not lifted. Should a semi-dry mix have been approved the means of ensuring that the semi-dry concrete does not lift during extraction of the casing shall be subject to approval.

5.142. Concrete level

When the casing is being extracted a sufficient quantity of concrete shall be maintained within it to ensure that pressure from external water or soil is exceeded and that the pile is neither reduced in section nor contaminated.

No concrete is to be placed in the boring once the bottom of the casing has been lifted above the top of the concrete; it shall be placed continuously as the casing is extracted until the desired head of concrete is obtained.

Adequate precautions shall be taken in all cases where excess hydraulic heads could be caused as the casing is withdrawn because of the displacement of water by the concrete as it flows into its final position against the walls of the shaft.

5.143. Vibrating extractors

The use of vibrating casing extractors will be permitted subject to clauses 1.111 and 1.112.

5.144. Water levels

In the event of the groundwater level being higher than the required pile head casting level shown on the Drawings, the Contractor shall submit his proposals for approval prior to placing concrete. The pile head shall not be left below the groundwater level unless approved precautions are taken.

5.15. TEMPORARY BACKFILLING ABOVE PILE CASTING LEVEL

After each pile has been cast any empty pile hole remaining shall be protected and shall be carefully backfilled as soon as possible with approved materials.

SECTION 6. STEEL PILES

6.01. *GENERAL*

All materials and work shall be in accordance with sections 1 and 2 of this specification, the Particular Specification and this section, except where there may be conflict of requirements, in which case those in the Particular Specification and this section shall take precedence.

This section does not include steel sheet piling.

6.02. *ORDERING OF PILES*

The Contractor shall seek the Engineer's instructions before ordering the piles. When preliminary piles are specified the instructions for the piles for the main work will not necessarily be given until the results of the driving and tests on the preliminary piles have been received and evaluated.

6.03. *PARTICULAR SPECIFICATION*

The following matters are, where appropriate, described in the Particular Specification

(*a*) instructions for ordering piles
(*b*) grades of steel
(*c*) sections of proprietary type of pile
(*d*) thickness of circumferential weld reinforcement
(*e*) minimum length of pile to be supplied
(*f*) types of head and toe preparation
(*g*) types of pile shoe
(*h*) types of coating
(*i*) thickness of primer and coats
(*j*) concreting of piles
(*k*) ownership of cut-off lengths of piles
(*l*) driving resistance
(*m*) penetration
(*n*) designed loads.

Materials

6.04. *PILE SHOES*

Cast steel pile shoes shall be of steel to BS 3100, grade A.

Welded fabricated pile shoes shall be to BS 4360, grade 43A.

6.05. *STRENGTHENING OF PILES*

The strengthening to the toe of a pile in lieu of a shoe or the strengthening of the head of a pile shall be made from material of the same grade as the pile unless otherwise approved.

92

6.06. *PILE SECTIONS AND PILE DIMENSIONS*

All piles shall be of the type and cross-sectional dimensions specified. For standard rolled sections the dimensional tolerances and weight shall comply with the relevant standard. For proprietary sections the dimensional tolerances shall comply with the manufacturer's standards. The rolling or manufacturing tolerances shall be such that the actual weight of sections does not differ from the theoretical weight by more than $-2\frac{1}{2}\%$ to $+5\%$ unless otherwise agreed.

For a tubular pile where the loads will be carried by the wall of the pile, and if the pile will be subjected to loads that induce reversal of stress during or after construction, the external dimensions at any section as measured by using a tape on the circumference shall not differ from the theoretical dimensions by more than -1% to $+1\%$. The rolling or manufacturing tolerances shall be such that the actual weight of any section does not differ from the theoretical weight by more than $-2\frac{1}{2}\%$ to $+5\%$.

For a tubular pile where the load will be static and will be carried by the wall of the pile or by a concrete core, the dimensions shall comply with API 5L. The rolling or manufacturing tolerances shall be such that the actual weight of any section does not differ from the theoretical weight by more than $-3\frac{1}{2}\%$ to $+10\%$.

6.07. *STRAIGHTNESS OF PILES*

For standard rolled sections the deviation from straightness shall not exceed $1\cdot04$ $(l-4\cdot5)$ where l is in metres and the deviation in millimetres. For proprietary sections made up from rolled sections the deviation from straightness shall not exceed $1/1000$ of the length of the pile.

For tubular piles the deviation from straightness shall not exceed $1/600$ of a length not exceeding 10 m. When two or more such lengths are joined the deviation from straightness shall not exceed $1/960$ of the completed length unless otherwise agreed by the Engineer.

6.08. *FABRICATION OF PILES*

The root edges or root faces of lengths of piles that are to be butt welded shall not differ by more than 25% of the thickness of piles not exceeding 12 mm thick or by more than 3 mm for piles thicker than 12 mm. When piles of unequal thickness are to be butt welded the thickness of the thinner material shall be the criterion.

Pile lengths shall be set up so that the differences in dimensions are matched as evenly as possible.

6.09. *MATCHING OF PILE LENGTHS*

Longitudinal seam welds and spiral seam welds of lengths of tubular piles forming a completed pile shall whenever possible be evenly staggered but if, in order to obtain a satisfactory match of the ends of piles or the specified straightness, the longitudinal seams or spiral seams are brought closely to one alignment at the joint then they shall be staggered by at least 100 mm.

6.10. WELDING

For a pile where the load will be carried by the wall or section of the pile, and if the pile will be subjected to loads that induce reversal of stresses during or after construction, the welding shall be to BS 5135.

For a tubular pile where the load will be static and will be carried by the wall of the pile or by a concrete core, the welding shall be to BS 2937.

6.11. INSPECTION AND TEST CERTIFICATES

The Contractor shall provide the Engineer with test certificates, analyses and mill sheets. The Contractor shall ensure that adequate notice be given to the Engineer when the processes can be inspected or tests can be witnessed.

The Engineer has the right to inspect and test at any stage of the manufacturing processes provided that, once he has been notified of when the materials will be ready for inspection, any delay in his attendance does not cause delay to or disrupt the production processes.

Workmanship

6.12. WELDERS' QUALIFICATIONS

Only welders who are qualified in the approved welding procedure in accordance with the tests laid down in the relevant British Standard shall be employed on the Permanent Works unless such work is in connection with the correction of minor surface defects. Copies of certificates relating to welders' tests shall be made available to the Engineer on request.

6.13. WELDING PROCEDURES

The Contractor shall submit for approval full details of the welding procedures and electrodes with drawings and schedules as may be necessary. Tests shall be undertaken as may be required by the relevant British Standard or as may be required by the Engineer.

6.14. ACCEPTANCE STANDARDS FOR RADIOGRAPHS OF WELDS

For a pile where the load will be carried by the wall or section of the pile, and if the pile will be subjected to loads that induce reversal of stress during or after construction, the acceptance standard for radiographs shall be in accordance with ASME Boiler and Pressure Vessel Code, section 8.

The thickness of welds shall be as specified in that code except that

94

the maximum permissible thickness of weld reinforcement for spiral or longitudinal welds shall not exceed 3·2 mm for wall thicknesses of 12·7 mm and less and 4·8 mm for wall thicknesses greater than 12·7 mm. For circumferential welds the same maximum thicknesses of weld reinforcement shall apply if so specified in clause 6.03(d).

For a pile where the loads will be static and will be carried by the wall of the pile, and where the pile will have a long length without lateral support, the acceptance standard for radiographs shall be in accordance with API 1104.

For a pile where the loads will be static and where the pile will receive lateral support, or where the load will be carried by a concrete core, radiographs will not be required unless specified, but welds shall be capable of withstanding handling, driving and design load stresses.

6.15. FABRICATION OF PILES ON SITE

When the pile lengths are to be made up on the Site all test procedures and dimensional tolerances shall conform to the Specification for the supply of pile material. Adequate facilities shall be provided for supporting and aligning the lengths of pile.

6.16. RADIOGRAPHIC TESTS

Radiographic tests shall be carried out as specified. While satisfactory results are being obtained, one radiograph 300 mm long shall be made for not less than 10% of the number of welded connections in the case of a pile where the load is carried by the wall or section of the pile, and for not greater than 10% of the number of welded connections in the case of a pile where the load is to be carried by a concrete core.

6.17. MARKING OF PILES

Each pile shall be clearly marked in white paint with its number and its overall length. In addition, each pile shall be marked at intervals of 250 mm along the top 3 m of its length before being driven.

6.18. LONGITUDINALLY WELDED PILES

6.181. Welded tube piles

Approval shall be obtained if different edge preparation from that shown on the Drawings is required for use with automatic welding machines or because of the method of rolling.

All welds shall be full penetration butt welds and, with the exception of continuous tube-making processes, longitudinal welds shall be made with extension plates at the starting and finishing points of each seam.

6.182. Welded box piles and proprietary sections

Welded box piles or proprietary sections made up from two or more hot rolled sections shall be welded in accordance with the manufacturer's standards.

95

6.183. Radiographs

During production of welded tube piles at least one radiograph approximately 300 mm long shall be required on each completed length as a spot check on weld quality. This shall be taken on a circumferential or longitudinal weld and its position shall be as directed by the Engineer.

The position and number of radiographs of welded box piles shall be as required.

6.184. Spirally welded piles

Prior to forming a spirally welded pile the edges of the strip shall be straight. The forming and welding process shall be approved. Before fabrication commences, tests as required by the Engineer in accordance with the standards for forming and welding shall be made to ensure that the welding procedure is satisfactory.

The Contractor shall satisfy the Engineer that all production welding is of sound quality. One of the following tests shall be carried out.

(a) For tubes of wall thickness 12 mm or less three spot check radiographs, one at each end of each length of the tube as manufactured and one at a position to be chosen at the time of testing by the Engineer, and spot check radiographs as required by the Engineer on the weld joints between strip lengths.

(b) For tubes of any wall thickness continuous ultrasonic examination over the whole weld, supplemented where necessary by radiographs to investigate defects revealed by the ultrasonic examination.

If the results of any weld test do not conform to the specified requirements, two additional specimens from the same length of pile shall be tested. In the case of failure of one or both of these additional tests the length of pile covered by the tests shall be rejected.

6.19. PILE COATINGS

6.191. Definition

The term 'coating' shall include the primer and the number of other coats specified.

6.192. Specialist labour

The preparation of surfaces and the application of the coats to form the coating shall be carried out by approved labour and, where appropriate, by specialist labour having experience in the preparation of the surface and the application of the coating specified.

6.193. Protection during coating

All work associated with surface preparation and coating shall be undertaken inside a substantially built waterproof structure unless such work is the lengthening of a partly driven pile.

6.194. Surface preparation

All surfaces to be coated shall be clean and dry and prepared by one or more of the following methods as specified.

(a) *Grit blasting.* All surfaces to be grit blasted shall be cleaned with an approved abrasive in accordance with either BS 4232 or SIS 05 59 00 to remove rust, mill scale and other adhering materials to provide a finish to second quality (near white) of BS 4232 or SIS 05 59 00, grade Sa $2\frac{1}{2}$, with a surface amplitude to 50–100 μm. The grit used shall not be manufactured from copper slag. Blast cleaning shall be done after fabrication unless otherwise approved. Unless an instantaneous recovery blasting machine is used the cleaned steel surface shall be air blasted with clean dry air, vacuum cleaned or otherwise freed from abrasive residues and dust immediately after cleaning. Fittings which cannot be blast cleaned satisfactorily may be cleaned as otherwise approved.

(b) *Pickling.* Pickling of piles shall be carried out by an approved process. Before being placed in the pickling solution all surfaces to be treated shall be thoroughly degreased and all paint marks removed. On completion the surfaces shall be given a neutralizing wash and the Contractor shall ensure that any by-product chemicals are compatible with the coating to be applied.

(c) *Flame cleaning.* The method of carrying out flame cleaning shall be approved, the number of passes and the speed of pass being determined by the apparatus used and the condition of the surface of the pile. On completion the surface shall be brushed to provide a finish to SIS 05 59 00, grade St 3. The priming or coating shall be applied while the surface is still warm.

(d) *Wire brushing.* All surfaces to be wire brushed shall be brushed to provide a finish to SIS 05 59 00, grade St 2.

(e) *Degreasing.* Degreasing with approved solvents compatible with the coating shall be carried out where necessary.

6.195. Application and type of primer

Immediately after surface preparation, the surface shall be coated with an approved primer or the specified coating to avoid recontamination. No coat shall be applied to a metal surface which is not thoroughly dry.

The primer shall be compatible with the specified coating and shall be such that if subsequent welding or cutting is to be carried out it shall not emit noxious fumes or be detrimental to the welding.

6.196. Control of humidity during coating

No coat shall be applied either to a cold surface or when the humidity in the vicinity of the surface is such that condensation would occur on the surface before or during the application or when the humidity could have an adverse effect on the coat.

When heating or ventilation is used to secure suitable conditions to allow coating to proceed, care shall be taken to ensure that heating or ventilation of a local surface does not have an adverse effect on adjacent surfaces or work already done.

6.197. Parts to be welded

The coating within 200 mm of a weld shall be applied after welding.

6.198. Thickness, number and colour of coats

The minimum thickness of the finished coating shall be as specified but in addition, where appropriate, the minimum thickness of each coat and the minimum number of coats that are to be applied shall also be as specified or as shown on the Drawings. Each coat shall be applied at an interval that ensures the proper hardening or curing of the previous coat and provides the specified dry film thickness without detriment to the surface finish.

Where more than one coat is applied to a surface each coat shall, if possible, be of a different colour from the previous coat. The colour of the final coat shall be approved.

6.199. Acceptability and inspection of coatings

The finished coating shall be generally smooth, of a dense and uniform texture and free from sharp protuberances or pin holes. Sags, dimpling or curtaining on up to 3% of the surface area of each member will be acceptable although sharp protuberances shall be removed by using a carpenter's chisel laid flat on the surface; areas from which material is removed shall be lightly wire brushed and recoated in accordance with the painting specification.

Any coat damaged by subsequent processes or which has deteriorated to an extent such that proper adhesion of the coating may not be obtained or maintained shall be recleaned to the original standard and recoated with the specified sequence of coats.

The completed coating shall be checked for thickness and continuity by an approved magnetic thickness gauge and Holiday detector. Areas where the thickness is less than that specified shall receive approved additional treatment. The completed coating shall also be checked for adhesion by driving a sharp carpenter's chisel laid almost flat through the coating and along the surface of the steel. A coating will be considered to be acceptably bonded if no separation is apparent between coats and if it can be seen to be adhering in the depressions on the exposed metal surface.

Alternatively the coatings may be checked by the cross-hatched method with each line spaced at ten times the thickness of the coating.

Tests shall be made for each 20 m² or part thereof of surface treated. Areas where the adhesion is not approved shall receive approved additional treatment. The coating shall be approved before pitching and driving of the piles.

6.20. HANDLING AND STORAGE OF PILES

All piles within a stack shall be in groups of the same length and on approved supports. All operations such as handling, transporting and pitching of piles shall be carried out in a manner such that damage to piles and their coatings is minimized.

6.21. DRIVING PILES

6.211. Leaders and trestles

At all stages during driving and until incorporation in the superstructure the pile shall be adequately supported and restrained by means of leaders, trestles, temporary supports or other guide arrangements to maintain position and alignment and to prevent buckling. These arrangements shall be such that damage to the piles or their coatings is minimized.

6.212. Performance of driving equipment

The Contractor shall satisfy the Engineer regarding the suitability, efficiency and energy of the driving equipment. Unless otherwise approved, drop hammers shall not be used from floating craft.

6.213. Length of piles

The length of pile to be driven in any position shall be approved.

The Contractor may, if approved, provide each pile in more than one length, the first length or subsequent lengths being extended during an interval in the pile driving operation. The extra lengths shall be cleaned and prepared to the tolerances in clause 6.06. During and after welding the lengths of pile shall be securely held to line and level.

6.214. Driving procedure and redrive checks

Each pile shall be driven continuously until the specified or approved set and/or depth has been reached, except that the Engineer may permit the suspension of driving if he is satisfied that the rate of penetration prior to the cessation of driving will be substantially reestablished on its resumption or if he is satisfied that the suspension of driving is beyond the control of the Contractor. A follower (long dolly) shall not be used unless approved, in which case the Engineer will require the set to be revised to take into account the reduction in the effectiveness of the hammer blow.

The Contractor shall inform the Engineer without delay if an unexpected change in driving characteristics is noted. A detailed record of the driving resistance over the full length of the next nearest available pile shall be taken if required.

At the start of work in a new area or section sets shall be taken at intervals during the last 3 m of the driving to establish the behaviour of the piles.

The Contractor shall give adequate notice and provide all facilities to enable the Engineer to check driving resistances. A set shall be taken only in the presence of the Engineer unless otherwise approved.

Redrive checks, if required, shall be carried out to an approved procedure.

6.215. Final set

The final set of each pile shall be recorded either as the penetration in millimetres per 10 blows or as the number of blows required to produce a penetration of 25 mm.

When a final set is being measured, the following requirements shall be met.

(*a*) The exposed part of the pile shall be in good condition without damage or distortion.

(*b*) The dolly and packing, if any, shall be in sound condition.

(*c*) The hammer blow shall be in line with the pile axis and the impact surfaces shall be flat and at right angles to the pile and hammer axis.

(*d*) The hammer shall be in good condition and operating correctly.

(*e*) The temporary compression of the pile shall be recorded if required.

6.216. Driving sequence and risen piles

Piles shall be driven in an approved sequence to minimize the detrimental effects of heave and lateral displacement of the ground.

When required, levels and measurements shall be taken to determine the movement of the ground or any pile resulting from the driving process.

When a pile has risen as a result of adjacent piles being driven, the Contractor shall submit to the Engineer his proposals for correcting this and the avoidance of it in subsequent work.

6.217. Preboring

If preboring is specified the pile shall be pitched into a hole prebored to the depth shown on the Drawings.

6.218. Jetting

Jetting shall be carried out only when the Contractor's detailed proposals have been approved, and not over the last 3 m of penetration.

6.22. PREPARATION OF PILE HEADS

If a steel superstructure is to be welded to piles, the piles shall be cut to within 10 mm of the levels shown on the Drawings. If piles are to be encased in concrete they shall be cut to within 20 mm of the levels shown on the Drawings, and protective coatings shall be removed from the surfaces of the pile heads down to a level 100 mm above the soffit of the concrete.

SECTION 7. TIMBER PILES

7.01. *GENERAL*

All materials and work shall be in accordance with sections 1 and 2 of this specification, the Particular Specification and this section, except where there may be conflict of requirements, in which case those in the Particular Specification and this section shall take precedence.

This section does not apply to piles for Temporary Works.

7.02. *PARTICULAR SPECIFICATION*

The following matters are, where appropriate, described in the Particular Specification

- (*a*) species and grades of timber
- (*b*) surface finishes to piles
- (*c*) length and dimensions
- (*d*) grades and types of pile shoe
- (*e*) driving resistance
- (*f*) penetration
- (*g*) designed loads
- (*h*) ownership of cut-off lengths of piles
- (*i*) preservative treatment.

Materials

7.03. *TIMBER*

7.031. *Species*

Timber shall be as specified or as otherwise approved.

7.032. *Grade*

The grade of the timber as described in CP 112, appendix A, shall be not less than 50.

7.033. *Sapwood*

Tree trunks for use as round piles shall have the bark removed but the sapwood left in place. They shall be treated with preservative. Sawn or hewn softwood or hardwood which is to be treated with preservative need not have the sapwood removed. Hardwood which is to be used untreated shall be free of sapwood.

7.034. *Tolerance in dimensions*

The dimensions of sawn piles shall be within the range of 6 mm less and 12 mm greater than their specified cross-sectional dimensions. The centroid of any cross-section of a sawn pile shall not deviate by more than 25 mm from the straight line connecting the centroids of the end faces of the pile.

Hewn piles may have a taper not exceeding 1 in 240 of their length;

at mid-length their cross-sectional dimensions shall be not more than 20 mm less than or greater than those specified. The centroid of any cross-section of a hewn pile shall not deviate by more than 40 mm from the straight line connecting the centroids of the end faces of the pile.

The minimum dimensions of a hewn pile shall not be less than the mean dimensions specified less 1/120 of the half length of the pile; from the dimensions so obtained, no deduction for the lower tolerance shall be made. For a round pile the minimum mid-length diameter, the maximum taper and maximum deviation from straight shall be as specified.

7.035. Condition

The timber shall be free from rot, pests, fungal or pest attack and from defects not permitted for its grade.

Timber to be treated with preservative shall have a moisture content of not more than that stated in BS 913 or in BS 4072. Timber not to be treated with preservative shall have a moisture content of not more than 23% at the time of installation, unless, before fungal growth can begin, it is placed in a permanently wet position.

7.04. PRESERVATIVES

Coal tar creosote shall comply with BS 144. Water-borne copper/chrome/arsenic composition shall comply with BS 4072, type 1 or type 2 as specified.

7.05. PILE SHOES

The material and dimensions of the pile shoes shall be as specified.

Cast iron shoes shall be made from chill hardened iron as used for making grey iron castings to BS 1452, grade 10. The chilled iron point shall be free from major blow holes and other surface defects.

Steel pile shoes shall be fabricated from steel to BS 3100, grade A.

Straps and other fastenings to cast pile shoes shall be of steel to BS 4360, grade 43A, and shall be cast into the point to form an integral part of the shoe.

Workmanship

7.06. INSPECTION AND STACKING

The Contractor shall notify the Engineer of the delivery of timber to the Site or to the place of preservative treatment, and provide all labour and materials to enable the Engineer to inspect each piece on all faces and to measure it at the time of unloading and immediately prior to driving.

Accepted timber shall be marked and stacked in lengths on paving

103

or drained hard ground. Each piece of timber shall be clear of the ground and have an air space around it. The baulks shall be separated by sticks or blocks placed vertically one above the other and closely spaced horizontally to avoid sagging of the timbers.

The timber shall be protected from the sun.

7.07. TREATMENT WITH PRESERVATIVE

Preservative treatment shall be carried out in accordance with BS 913 or BS 4072 as specified. Cutting and boring of timber shall be done as far as possible before preservative treatment, but where this is impracticable all surfaces subsequently cut or bored shall be heavily coated with preservative as specified in the relevant British Standard for preservative treatment.

7.08. PILE SHOES

The shoe shall be attached to the pile by steel straps fixed, spiked, screwed or bolted to the timber. The shoe shall be co-axial with the pile and firmly bedded to it.

7.09. PILE HEADS

The pile head shall be flat and at right angles to the axis of the pile.

Unless otherwise specified the head of each pile shall be trimmed to a round cross-section and fitted with a tight steel ring. The ring shall be not less than 50 mm by 20 mm in cross-section and the join shall be welded for its full section. The external diameter of the ring shall be that of the least transverse dimension of the head of the pile. The top of the ring shall be between 10 mm and 20 mm from the top of the pile. If the ring is displaced during driving it shall be refitted. If the ring is broken a new ring shall be fitted.

As an alternative to a ring, a metal helmet may be used, the top of the pile being trimmed to fit closely into the recess of the underside of the helmet. A hardwood dolly and, if necessary, a packing shall be used above the helmet.

If during driving the head of the pile becomes excessively broomed or otherwise damaged, the damaged part shall be cut off, the head retrimmed and the ring or helmet, if any, refitted.

7.10. SPLICING AND SCARFING

Piles shall be provided in one piece unless otherwise approved. A splice shall be capable of resisting safely any stresses which may develop during lifting, pitching or driving, and under the designed working load. The position and details of the splice shall be subject to approval.

A splice shall be made in accordance with the following principles or by another approved method. The two timbers shall be of the same sectional dimensions and each cut at right angles to its axis to make contact over the whole of the cross-section when the two timbers are co-axial. An approved jointing compound shall be used at the contact surface. The two timbers shall be joined by a steel tube of round or rectangular section to fit the timbers closely. The tube shall be bolted, screwed or spiked to the timbers to keep the joined ends in close contact.

Where it is necessary to extend a partly-driven pile, the upper part must be securely supported during the making of the joint.

A scarf may be formed to details to be approved.

7.11. DRIVING PILES

7.111. Leaders and trestles

At all stages during driving and until incorporation in the superstructure the pile shall be adequately supported and restrained by means of leaders, trestles, temporary supports or other guide arrangements to maintain the position and alignment and to prevent bending. These arrangements shall be such that damage to the piles or the preservative treatment does not occur.

7.112. Performance of driving equipment

The Contractor shall satisfy the Engineer regarding the suitability, efficiency and energy of the driving equipment. Unless otherwise approved, drop hammers shall not be used from floating craft.

7.113. Length of piles

The length of pile to be driven in any position shall be approved.

7.114. Driving procedure and redrive checks

Each pile shall be driven continuously until the specified or approved set and/or depth has been reached, except that the Engineer may permit the suspension of driving if he is satisfied that the rate of penetration prior to the cessation of driving will be substantially re-established on its resumption or if he is satisfied that the suspension of driving is beyond the control of the Contractor. A follower (long dolly) shall not be used unless approved, in which case the Engineer will require the set to be revised to take into account the reduction in the effectiveness of the hammer blow.

The Contractor shall inform the Engineer without delay if an unexpected change in driving characteristics is noted. A detailed record of the driving resistance over the full length of the next nearest available pile shall be taken if required.

At the start of work and in a new area or section sets shall be taken at intervals during the last 3 m of the driving to establish the behaviour of the piles.

The Contractor shall give adequate notice and provide all facilities to enable the Engineer to check driving resistances. A set shall be taken only in the presence of the Engineer unless otherwise approved. Redrive checks, if required, shall be carried out to an approved procedure.

7.115. *Final set*

The final set of each pile shall be recorded either as the penetration in millimetres per 10 blows or as the number of blows required to produce a penetration of 25 mm.

When a final set is being measured, the following requirements shall be met.

(*a*) The exposed part of the pile shall be in good condition without damage or distortion.

(*b*) The dolly and packing, if any, shall be in sound condition.

(*c*) The hammer blow shall be in line with the pile axis and the impact surfaces shall be flat and at right angles to the pile and hammer axis.

(*d*) The hammer shall be in good condition and operating correctly.

(*e*) The temporary compression of the pile shall be recorded if required.

7.116. *Spliced or scarfed piles*

Spliced or scarfed piles shall be observed continuously during driving to detect any departure from true alignment of the two parts. If any such departure occurs, driving shall be suspended and the Engineer shall be informed.

7.117. *Driving sequence and risen piles*

Piles shall be driven in an approved sequence to minimize the detrimental effects of heave and lateral displacement of the ground.

When required, levels and measurements shall be taken to determine the movement of the ground or any pile resulting from the driving process.

When a pile has risen as a result of adjacent piles being driven the Contractor shall submit to the Engineer his proposals for correcting this and the avoidance of it in subsequent work.

7.118. *Preboring*

If preboring is specified the pile shall be pitched into a hole prebored to the depth shown on the Drawings.

7.119. *Jetting*

Jetting shall be carried out only when the Contractor's detailed proposals have been approved, and not over the last 3 m of penetration.

7.12. PREPARATION OF PILE HEADS

After driving the piles shall be cut off square at the designed cut-off level and the cut surfaces shall be heavily coated with preservative as specified for the initial treatment.

SECTION 8. PILE TESTING

8.01. GENERAL

All materials and work shall be in accordance with section 1 and those sections of this specification relating to the manufacture and installation of the type of pile to be tested, the Particular Specification and this section, except where there may be conflict of requirements, in which case those in the Particular Specification and in this section shall take precedence.

This section deals with the testing of a pile by the application of an axial load or force. It covers vertical and raking piles tested in compression (i.e. subjected to loads or forces in a direction such as would cause the piles to penetrate further into the ground) and vertical or raking piles tested in tension (i.e. subjected to forces in a direction such as would cause the piles to be extracted from the ground).

8.02. PARTICULAR SPECIFICATION

The following matters are, where appropriate, described in the Particular Specification
- (*a*) type of pile
- (*b*) preliminary pile tests and test loads
- (*c*) working pile tests and proof loads
- (*d*) type of tests
- (*e*) removal of Temporary Works and equipment
- (*f*) penetration
- (*g*) driving resistance.

8.03. DEFINITIONS

Allowable load: the load which may be safely applied to a pile after taking into account its ultimate bearing capacity, negative friction, pile spacing, overall bearing capacity of the ground below and allowable settlement.

Compression pile: a pile which is designed to resist an axial force such as would cause it to penetrate further into the ground.

Constant rate of penetration (CRP) test: a test in which the pile is made to penetrate the soil from its position as installed at a constant speed while the force applied at the top of the pile to maintain the rate of penetration is continuously measured. The force/penetration relationship obtained does not represent an equilibrium condition between load and settlement.

Constant rate of uplift (CRU) test: a test in which the pile is extracted from its position as installed in the soil at a constant speed while the force applied at the top of the pile to maintain the rate of uplift is continuously measured. The force/uplift relationship obtained does not represent an equilibrium condition between the extractive force and the uplift.

Kentledge: the dead weight used in a loading test.

Maintained load test: a loading test in which each increment of load is held constant either for a defined period of time or until the rate of movement (settlement or uplift) falls to a specified value.

Preliminary pile: a pile installed before the commencement of the main piling works or a specific part of the Works for the purpose of establishing the suitability of the chosen type of pile and for confirming its design, dimensions and bearing capacity.

Proof load: a load applied to a selected working pile to confirm that it is suitable for the load at the settlement specified. A proof load should not normally exceed 150% of the working load on a pile.

Reaction system: the arrangement of kentledge, piles, anchors or rafts that provides a resistance against which the pile is tested.

Raking pile: a pile installed at an inclination to the vertical.

Tension pile: a pile which is designed to resist an axial force such as would cause it to be extracted from the ground.

Test pile: any pile to which a test is, or is to be, applied.

Ultimate bearing capacity: the load at which the resistance of the soil becomes fully mobilized.

Working load: the load which the pile is designed to carry.

Working pile: one of the piles forming the foundation of a structure.

8.04. SUPERVISION

All tests shall be carried out only under the direction of an experienced and competent supervisor conversant with the test equipment and test procedure. All personnel operating the test equipment shall have been trained in its use.

8.05. SAFETY PRECAUTIONS
8.051. General

When preparing for, conducting and dismantling a pile test the Contractor shall carry out the requirements of the various Acts, orders, regulations and other statutory instruments that are applicable to the work for the provision and maintenance of safe working conditions, and shall in addition make such other provision as may be necessary to safeguard against any hazards that are involved in the testing or preparations for testing.

8.052. Kentledge

Where kentledge is used the Contractor shall construct the foundations for the kentledge and any cribwork, beams or other supporting structure in such a manner that there will not be differential settlement, bending or deflexion of an amount that constitutes a hazard to safety or impairs the efficiency of the operation. The kentledge shall

109

be adequately bonded, tied or otherwise held together to prevent it falling apart, or becoming unstable because of deflexion of the supports.

The weight of kentledge shall be greater than the maximum test load and if the weight is estimated from the density and volume of the constituent materials an adequate factor of safety against error shall be allowed.

8.053. Tension piles and ground anchors

Where tension piles or ground anchors are used the Contractor shall ensure that the load is correctly transmitted to all the tie rods or bolts. The extension of rods by welding shall not be permitted unless it is known that the steel will not be reduced in strength by welding. The bond stresses of the rods in tension shall not exceed normal permissible bond stresses for the type of steel and grade of concrete used.

8.054. Testing equipment

In all cases the Contractor shall ensure that when the hydraulic jack and load measuring device are mounted on the pile head the whole system will be stable up to the maximum load to be applied. Means shall be provided to enable dial gauges to be read from a position clear of the kentledge stack or test frame in conditions where failure in any part of the system due to overloading, buckling, loss of hydraulic pressure and so on might constitute a hazard to personnel.

The hydraulic jack, pump, hoses, pipes, couplings and other apparatus to be operated under hydraulic pressure shall be capable of withstanding a test pressure of $1\frac{1}{2}$ times the maximum working pressure without leaking.

The maximum test load or test pressure expressed as a reading on the gauge in use shall be displayed and all operators shall be made aware of this limit.

8.06. CONSTRUCTION OF A PRELIMINARY PILE TO BE TESTED

8.061. Notice of construction

The Contractor shall give the Engineer at least 48 hours' notice of the commencement of construction of any preliminary pile which is to be test loaded.

8.062. Method of construction

Each preliminary test pile shall be constructed in a manner similar to that to be used for the construction of the working piles, and by the use of similar equipment and materials. Any variation will only be permitted with prior approval.

Extra reinforcement and concrete of increased strength will be permitted in the shafts of preliminary piles at the discretion of the Engineer.

8.063. Boring or driving record

For each preliminary pile which is to be tested a detailed record of the soils encountered during boring, or of the progress during driving shall be made and submitted to the Engineer daily not later than noon on the next working day.

Where the Engineer requires soil samples to be taken or in situ tests to be made in bored piles, the Contractor shall give the results of such tests to the Engineer without delay.

8.064. Cut-off level

The pile shaft shall terminate at the normal cut-off level or at a level required by the Engineer.

The pile shaft shall be extended where necessary above the cut-off level of working piles so that gauges and other apparatus to be used in the testing process will not be damaged by water or falling debris.

Where the pile shaft is extended above the cut-off level of a working pile in a soil which would influence the load bearing capacity of the pile, a sleeve shall be installed and kept in place during testing to eliminate friction which would not arise in the working pile. Alternatively, if the friction above the designed cut-off level can be calculated with reasonable accuracy, with the approval of the Engineer a sleeve need not be used, but the calculated friction must be taken into account in assessing the load being applied to the pile.

If the cut-off level is below ground level, the shaft is not extended and there is a risk of the borehole collapsing, a sleeve shall be left in place or inserted above the pile shaft, or other approved action shall be taken. Adequate clearance shall be given between the top of the pile shaft and the bottom of the sleeve to permit unrestricted movement of the pile.

8.065. Pile head for compression test

For a pile that is tested in compression, the pile head or cap shall be formed to give a plane surface which is normal to the axis of the pile, sufficiently large to accommodate the loading and settlement-measuring equipment and adequately reinforced or protected to prevent damage from the concentrated application of load from the loading equipment.

The pile cap shall be concentric with the test pile; the joint between the cap and the pile shall have a strength equivalent to that of the pile.

Sufficient clear space shall be made under any part of the cap projecting beyond the section of the pile so that, at the maximum expected settlement, load is not transmitted to the ground except through the pile.

111

8.066. *Pile connection for tension test*

For a pile that is tested in tension, means shall be provided for transmitting the test load axially to the pile.

The connection between the pile and the loading equipment shall be constructed in such a manner as to provide a strength equal to the maximum load which is to be applied to the pile during the test with an appropriate factor of safety on the structural design.

8.07. PREPARATION OF A WORKING PILE TO BE TESTED

If a test is required on a working pile the Contractor shall cut down or otherwise prepare the pile for testing as required by the Engineer in accordance with clauses 8.064, 8.065 and 8.066.

8.08. CONCRETE TEST CUBES

Three test cubes shall be made from the concrete used in the preliminary test pile and from the concrete used for building up a working pile. If a concrete cap is cast separately from a preliminary pile or a working pile a further three cubes shall be made from this concrete. The cubes shall be made and tested in accordance with BS 1881.

The pile test shall not be started until the strength of the cubes taken from the pile exceeds twice the average direct stress in any pile section under the maximum required test load and the strength of the cubes taken from the cap exceeds twice the average stress at any point in the cap under the same load. Variation of procedure will be permitted only if approved.

8.09. REACTION SYSTEMS
8.091. *Compression tests*

Compression tests shall be carried out using kentledge, tension piles or specially constructed anchorages. Kentledge shall not be used for tests on raking piles.

Where kentledge is to be used, it shall be supported on cribwork disposed around the pile head so that its centre of gravity is on the axis of the pile. The bearing pressure under supporting cribs shall be such as to ensure stability of the kentledge stack. Kentledge shall not be carried directly on the pile head, except when directed by the Engineer.

8.092. *Tension tests*

Tension tests shall be carried out using compression piles or rafts constructed on the ground. The use of inclined reaction piles, anchors or rafts is not precluded, subject to approval. In all cases the resultant force of the reaction system shall be co-axial with the test pile.

8.093. *Working piles*

Working piles shall not be used as reaction piles without approval.

Where working piles are used as reaction piles their movement shall be measured to within an accuracy of 0·5 mm.

8.094. *Spacing*

Where kentledge is used for loading vertical piles in compression, the distance from the edge of the test pile to the nearest part of the crib supporting the kentledge stack in contact with the ground shall be not less than 1·3 m.

The centre to centre spacing of vertical reaction piles, including working piles used as reaction piles, from a test pile shall be not less than three times the diameter of the test pile or the reaction piles or 2 m, whichever is the greatest. Where a pile to be tested has an enlarged base, the same criterion shall apply with regard to the pile shaft, with the additional requirement that the surface of no reaction pile shall be closer to the base of the test pile than one half of the enlarged base diameter.

Where ground anchors are used to provide a test reaction for loading in compression, no part of the section of the anchor transferring load to the ground shall be closer to the test pile than three times the diameter of the test pile. Where the pile to be tested has an enlarged base, the same criterion shall apply with regard to the pile shaft, with the additional requirement that no section of the anchor transferring load to the ground shall be closer to the pile base than a distance equal to the base diameter.

8.095. *Adequate reaction*

The size, length and number of the piles or anchors, or the area of the rafts, shall be adequate to transmit the maximum test load to the ground in a safe manner without excessive movement or influence on the test pile.

8.096. *Care of piles*

The method employed in the installation of any reaction piles, anchors or rafts shall be such as to prevent damage to any test pile or working pile.

8.097. *Loading arrangement*

The loading arrangement used shall be designed to transfer safely to the test pile the maximum load required in testing. Full details shall be submitted to the Engineer prior to any work related to the testing process being carried out on the Site.

8.10. *EQUIPMENT FOR APPLYING LOAD*

The equipment used for applying load shall consist of one or more hydraulic rams or jacks. The total capacity of the jacks shall be at least equal to the required maximum load. The jack or jacks shall be arranged in conjunction with the reaction system to deliver an axial load to the test pile. The complete system shall be capable of transferring the maximum load required for the test.

8.11. MEASUREMENT OF LOAD

The load shall be measured by a load measuring device and by a calibrated pressure gauge included in the hydraulic system. Readings of both the load measuring device and the pressure gauge shall be recorded. In interpreting the test data the values given by the load measuring device shall normally be used; the pressure gauge readings are required as a check for gross error.

The load measuring device may consist of a proving ring, load measuring column, pressure cell or other appropriate system. A spherical seating shall be used in conjunction with any devices that are sensitive to eccentric loading; care must be taken to avoid any risk of buckling. Load measuring devices and jacks shall be short in axial length in order to achieve the best possible stability; the Contractor shall pay attention to details in order to ensure that axial loading is maintained.

The load measuring device shall be calibrated before and after each series of tests, whenever adjustments are made to the device or at intervals appropriate to the type of equipment. The pressure gauge and hydraulic jack shall be calibrated together. Certificates of calibration shall be supplied to the Engineer.

The Engineer's agreement shall be obtained in writing before any modification of this procedure is adopted.

8.12. ADJUSTABILITY OF LOADING EQUIPMENT

The loading equipment shall be capable of adjustment throughout the test to obtain a smooth increase of load or to maintain each load constant at the required stages of a maintained loading test.

8.13. MEASURING MOVEMENT OF PILE HEADS
8.131. Maintained load test

In a maintained load test movement of the pile head shall be measured by one of the methods in clauses 8.133, 8.134, 8.135 and 8.136 in the case of vertical piles, or by one of the methods in clauses 8.134, 8.135 and 8.136 in the case of raking piles, as required.

8.132. CRP and CRU tests

In a CRP or a CRU test the method in clause 8.164 shall be used. Check levelling of the reference frame or on the pile head shall not be required. The dial gauge shall be graduated in divisions of 0·02 mm or less.

8.133. Levelling method

An optical or any other levelling method by reference to an external datum may be used.

Where a level and staff are used, the level and scale of the staff shall be chosen to enable readings to be made to within an accuracy of 0·5 mm. A scale attached to the pile or pile cap may be used instead of a levelling staff. At least two datum points shall be established on permanent objects or other well-founded structures or deep datum points shall be installed. Each datum point shall be situated so that only one setting up of the level is needed.

No datum point shall be affected by the test loading or other operations on the Site.

Where another method of levelling is proposed this shall be approved in writing.

8.134. Independent reference frame

An independent reference frame may be set up to permit measurement of the movement of the pile. The supports for the frame shall be founded in such a manner and at such a distance from the test pile, kentledge support cribs, reaction piles, anchorages and rafts that movements of the ground in the vicinity of the equipment do not cause movement of the reference frame during the test which will affect the required accuracy of the test. Check observations of any movements of the reference frame shall be made and a check shall be made of the movement of the pile head relative to an external datum during the progress of the test. In no case shall the supports be less than three test pile diameters or 2 m, whichever is the greater, from the centre of the test pile.

The measurement of pile movement shall be made by two dial gauges rigidly mounted on the reference frame that bear on surfaces normal to the pile axis fixed to the pile cap or head. Alternatively the gauges may be fixed to the pile and bear on surfaces on the reference frame. The dial gauges shall be placed in diametrically opposed positions and be equidistant from the pile axis. The dial gauges shall enable readings to be made to within an accuracy of 0·1 mm.

The reference frame shall be protected from sun and wind.

8.135. Reference wire

A reference wire shall be held under constant tension between two foundations formed as in the method in clause 8.134. The wire shall be positioned against a scale fixed to the pile and the movement of the scale relative to the wire shall be determined.

Check observations of any movements of the supports of the wire shall be made or a check shall be made of the movement of the pile head as in the method in clause 8.133. Readings shall be taken to within an accuracy of 0·5 mm.

The reference wire shall be protected from sun and wind.

8.136. Other methods

The Contractor may submit any other method for measuring the movement of pile heads for approval.

8.14. PROTECTION OF TESTING EQUIPMENT

8.141. Protection from weather

Throughout the test period all equipment for measuring load and movement shall be protected from the weather.

8.142. Prevention of disturbance

Construction equipment and persons who are not involved in the testing process shall be kept at a sufficient distance from the test to avoid disturbance to the measurement apparatus.

8.15. SUPERVISION

8.151. Notice of test

The Contractor shall give the Engineer at least 24 hours' notice of the commencement of the test.

8.152. Records

During the progress of a test, the testing equipment and all records of the test as required in clause 8.172 shall be available for inspection by the Engineer.

8.16. TEST PROCEDURE

8.161. Proof test by maintained load test

The maximum load which shall be applied in a proof test on a working pile is $1\frac{1}{2}$ times the working load. The loading and unloading shall be carried out in stages as shown in Table 8.

Following each application of an increment of load the load shall be held for not less than the period shown in Table 8 or until the rate of settlement is less than 0·25 mm/h and slowing down. The rate of settlement shall be calculated from the slope of the curve obtained by plotting values of settlement versus time and drawing a smooth curve through the points.

Each stage of unloading shall proceed after the expiry of the period shown in Table 8.

For any period when the load is constant, time and settlement shall be recorded immediately on reaching the load and at approximately 15 min intervals for 1 h, at 30 min intervals between 1 h and 4 h, and at 1 h intervals between 4 h and 12 h after the application of the increment of load.

Table 8

Load, percentage of working load	Minimum time of holding load
25	1 h
50	1 h
75	1 h
100	1 h
75	10 min
50	10 min
25	10 min
0	1 h
100	6 h
125	1 h
150	6 h
125	10 min
100	10 min
75	10 min
50	10 min
25	10 min
0	1 h

8.162. Use of CRP test

The ultimate bearing capacity of preliminary piles and piles which are not to be used subsequently as working piles shall be determined by the CRP test, unless otherwise required by the Engineer.

8.163. Use of CRU test

The ultimate capacity in tension of preliminary piles and piles which are not to be used subsequently as working piles shall be determined by the CRU test, unless otherwise required by the Engineer.

8.164. Procedure for a CRP or CRU test

The rate of loading shall be such that a CRP or CRU is maintained throughout the test insofar as is practicable. The rate of movement of each pile to be tested shall be agreed with the Engineer prior to the start of the test.

Readings of load, penetration or uplift, and time shall be made simultaneously at regular intervals; the interval chosen shall be such that a curve of load versus penetration or uplift can be plotted without ambiguity.

Loading shall be continued until one of the following results is obtained

117

(*a*) the maximum required test load in clause 8.02 is reached

(*b*) a constant or reducing load has been recorded for an interval of penetration or uplift of 10 mm

(*c*) a total movement of the pile base equal to 10% of the base diameter, or any other greater value of movement required, has been reached.

The load shall then be reduced in five approximately equal stages to zero load, penetration or uplift and load at each stage and at zero load being recorded.

8.165. Combined proof test and CRP or CRU test

Where required a proof test by maintained loading shall be carried out prior to a CRP or CRU test.

8.17. PRESENTATION OF RESULTS
8.171. Results to be submitted

Results shall be submitted as

(*a*) a summary in writing to the Engineer, unless otherwise directed within 24 hours of the completion of the test, which shall give

(i) for a proof test by maintained load for each stage of loading, the period for which the load was held, the load and the maximum settlement or uplift recorded

(ii) for a CRP or CRU test the maximum load reached and a graph of load against penetration or load against uplift

(*b*) the completed schedule of recorded data as in clause 8.172 within seven days of the completion of the test.

8.172. Schedule of recorded data

The Contractor shall provide information about the tested pile in accordance with the following schedule where applicable.

(*a*) *General*

Site location

Contract identification

Proposed structure

Main Contractor

Piling Contractor

Engineer

Client

Date of test

(*b*) *Pile details*

(i) *All types of pile*

Identification (number and location)

Position relative to adjacent piles

Brief description of location (e.g. in cofferdam, in cutting, over water)

Ground level at pile position

Head level at which test load is applied

Type of pile (e.g. precast reinforced concrete, steel H, bored in place, driven in place, composite type)

Vertical or raking, compression or tension

Shape and size of cross-section of pile, position of change in cross-section

Shoe or base details

Head details

Length in ground

Level of toe

Any permanent casing or core

(ii) *Concrete piles*

Concrete mix

Aggregate type and source

Cement type

Slump

Cube test results for pile and cap

Date of casting of precast pile

Reinforcement

(iii) *Steel piles*

Steel quality

Coating

Filling

(c) *Installation details*

  (i) *All piles*

Dates and times of boring, driving and concreting of test pile and adjacent piles

Unexpected circumstances and difficulties

Date and time of casting concrete pile cap

Start and finish of each operation during driving or installation of a pile and subsequent testing

Difficulties in handling, pitching and driving pile

Delays due to sea and weather conditions

  (ii) *Bored piles*

Type of equipment used and method of boring

Temporary casing, method of installation and extraction

Strata encountered during boring

Water encountered during boring

Method of placing concrete and conditions pertaining

Volume of concrete placed

Concrete level before and after extraction of casing

(iii) *Driven preformed piles and piles driven in place*

Method of support of pile and hammer (frame, hanging leaders, suspended hammer or other method)

Driven length of pile or temporary casing at final set

Hammer type, size or weight

Dolly and packing, type and condition before and after driving

Driving log (depth, blows per 250 mm, interruptions or breaks in driving)

Final set in number of blows to produce penetration of 25 mm

Redrive check, time interval and set in number of blows to produce penetration of 25 mm

At final set and at redrive set, for drop or single acting hammer the length of the drop or stroke, for diesel hammer the length of the stroke and the blows per minute, for double-acting hammer the number of blows per minute

Condition of pile head or temporary casing after driving

Use of a follower

Use of preboring

Use of jetting

Lengthening

Details of temporary casing

Concrete level before and after extraction of casing

Method of placing concrete and conditions pertaining

(*d*) *Test procedure*

Weight of kentledge

Tension pile, ground anchor or compression pile details

Plan of test arrangement showing position and distances of kentledge supports, rafts, tension or compression piles and reference frame to test pile

Jack capacity

Method of load measurement

Method(s) of penetration or uplift measurement

Proof test by maintained loading and CRP or CRU

Relevant dates and times

(*e*) *Test results*

In tabular form

In graphical form: load plotted against settlement, load plotted against uplift, with times

Ground heave

Effect on adjacent structure

(*f*) *Site investigation*

Site investigation report number

Borehole references

8.18. COMPLETION OF A TEST

8.181. Measuring equipment

On completion of a test all equipment and measuring devices shall be dismantled, checked and either stored so that they are available for use in further tests or removed from the Site as specified.

8.182. Kentledge

Kentledge and its supporting structure shall be removed from the test pile and stored so that they are available for use in further tests or removed from the Site as specified.

8.183. Ground anchors and temporary piles

On completion of a preliminary test, tension piles and ground anchors shall be cut off below ground level, removed from the Site and the ground made good with approved material as specified.

8.184. Preliminary test pile cap

The pile cap, if formed in concrete, shall be broken off and the resulting material disposed of off the Site. If the pile cap is made of steel it shall be cut off and stored so that it is available for use in further tests or removed from the Site as specified.

The pile head shall be made good or extended to the cut off level in clause 8.064.

8.185. Proof test pile cap

On completion of a test on a proof pile, the test pile cap, if in concrete, shall be stripped and left in a state ready for incorporation in the Permanent Works and the resulting material disposed of off the Site.

If the pile cap is made of steel it shall be cut off and stored so that it is available for use in further tests or removed from the Site as specified.

8.186. Ground anchors and temporary piles

On completion of a proof test, temporary piles and ground anchors shall be removed, cut off as specified or, if approved, incorporated in the Permanent Works.

MODEL PROCEDURES FOR SITE INVESTIGATION FOR PILING

These model procedures are for guidance only and are to be excluded from contract documents

GENERAL

The report *Placing and management of contracts for building and civil engineering work,** generally referred to as the Banwell report, emphasizes the importance of establishing clearly, before a contract is placed, the obligations which the Contractor is to undertake, the time allowed for the work and the system of payment. For a site investigation the Client or Engineer inviting tenders must convey clearly to tenderers the nature of the results he requires.

SI 2. Paragraphs SI 7–SI 59 are intended to assist in the preparation of a contract and specification for site investigation work which is initiated by the Engineer to enable him to design the most suitable foundation for a particular site. When undertaking the preliminary and subsequent stages of investigation the Engineer should ensure that the information obtained covers the requirement for piling where this needs to be considered among the possible solutions to foundation design problems on the Site.

SI 3. In the model specification for site investigation for piling the 1957 edition of CP 2001 is referred to as CP 2001/57 to distinguish it from later revisions.

TERMINOLOGY

SI 4. All concerned with the framing and use of specifications for site investigations should be familiar with and have access to the latest current version of CP 2001. In these procedures and in the model specification the word 'Contractor' means the organization undertaking the site investigation.

SI 5. Rocks are described in the ranges from 'very weak' to 'extremely strong' as set out in a report of the Working Party of the Engineering Group of the Geological Society.†

SI 6. The terms 'soft' and 'hard' are not used to describe rock because they are applicable only to cohesive soils.

---

* HMSO, London, 1964.

† Geological Society Engineering Group Working Party Report. The logging of rock cores for engineering purposes. *Q. J. Engng Geol.,* 1970, **3,** No. 1, 1–24.

PARTICULAR CLAUSES

SI 7. It is by no means always desirable that the items in section SI 1 should have full descriptions. Only confusion will result from the inclusion of information or requirements which the Engineer does not need or on which the Contractor does not have to take action. Once an exact definition of the site investigation has been made, it will be possible to recognize the items that require elaboration.

SI 8. For most site investigations it will be necessary for the Engineer to provide in addition to the Specification an addendum to the Particular Specification detailing matters pertinent to the particular site. Particular information will be needed about clauses SI 1.01 and SI 1.02 as follows.

ACCESS TO THE SITE (clause SI 1.01(d))

SI 9. A statement should be made as to whether or not the Client owns the Site. Where he does not the Contractor should be given a copy of the letter or other document authorizing access to the Site. It is desirable for this be included in the tender documents.

MARINE BORINGS (clause SI 1.01(g))

SI 10. Marine boring is usually difficult. Its costs are greatly affected by weather, marine conditions and hours of daylight. Apart from exploring the ground, reasons for breakdown of craft and for standby time are essential as part of the report on the work. Records should include accounts of weather, wind, waves, swell (magnitude and direction), such matters as dragging anchors or loss of buoys, currents and all operations concerning boring tubes, including incidents of vibration, oscillation and nature of damage or breakage.

SI 11. Clauses should be given for the priority afforded to passage of ships, the employment of a diver and the removal of broken casings and sunken craft.

WHARFAGE AND BERTHING FACILITIES (clause SI 1.01(h))

SI 12. The wharf or berth for use by the Contractor's craft and access to it should be defined.

OTHER MARINE WORK

SI 13. Other clauses covering particular aspects of marine site investigation may be required in the Specification.

TYPE OF INVESTIGATION AND REPORT REQUIRED (clause SI 1.02(a))

SI 14. It may be specified that only a factual report comprising the borehole records and the results of in situ and laboratory tests is required. Alternatively, the Contractor may be required to interpret the field and laboratory data to provide a report. In its simplest form this might consist only of recommendations for soil strength or density parameters from which piling contractors may judge skin friction or end bearing values. At the other extreme the Contractor may be required to provide a report in which the relative merits of various forms of foundation are discussed, e.g. comparison of rafts with piled foundations, the suitability of various piling systems, or the effects of grouping, penetration depths and base enlargement on bearing capacity and settlement. Such a comprehensive report would involve the Contractor in detailed calculations varying in complexity with the ground conditions.

SI 15.   The scope of the Contractor's report and the work and time taken to prepare it will in many cases be impossible to assess until the results of boring and testing are available. The extent of work for providing simple skin friction and end bearing values can vary widely, depending on the type of soil and multiplicity of soil layers. Therefore it is recommended that the Contract should provide for discussion on the scope of the report when the preliminary results of the site investigations are available, and that the form of contract should allow the Contractor to be reimbursed on a time basis for all report work other than that involved in preparing factual details such as borehole records and test data.

SI 16.   Possible requirements are

(a) the results of specified observations and tests only

(b) the results of specified observations and tests together with an interpretation of the results as they affect certain stated design problems, e.g. triaxial tests on clay may be required with an opinion on the strength parameters for estimating the ultimate loads for driven or in situ piles penetrating this clay

(c) the results of specified observations and tests together with interpretations of the results, calculations for certain postulated forms of construction and recommendations for a preferred design solution.

RESPONSIBILITY (clauses SI 1.02(a) and SI 1.02(e))

SI 17.   It should be specified that when calculations, interpretations and recommendations are made under clause SI 1.02(a) (see paragraph SI 16(c)), the names, qualifications and details of the experience of those who are responsible for them, and their positions within the Contractor's organization, must be given.

BOREHOLES (clause SI 1.02(b))

SI 18.   The Contractor should be given the opportunity to comment on the number, depths and positions of the boreholes and invited to suggest alterations and additions both before and during the execution of the work. This is particularly important where he is required to interpret the results.

EXISTING SOURCES OF POLLUTION

SI 19.   Enquiries should be made to ascertain any natural or man-made sources of pollution which might deleteriously affect the ground-water.

BORING AND SAMPLING

BORING METHODS (clause SI 2.01)

SI 20.   Boring by shell and auger* or rotary power auger equipment in very weak rock is suitable only for determining the level of the interface between the rock head and overlying soils. The equipment is unsuitable for establishing a suitable level at which piles can be terminated. For example, it is frequently impossible to distinguish the level at which there is a change from very weak weathered rock to stronger relatively unweathered rock. This is because, in the presence of water, shell and auger or rotary power auger drilling reduces rock to a sludge. The strength of rock or the degree of cementation of the rock particles cannot be assessed with any reliability. In situ penetration testing by means of the standard penetration test (clause SI 3.02) or the dynamic cone penetration test (clause SI 3.04) can be used in conjunction with shell and auger or rotary power auger drilling to obtain quantitative information on the relative strength and other physical characteristics of weathered rocks.

SI 21.   The best method of establishing a suitable level for the termination of piles is by rotary core drilling with the core diameter large enough to permit the extraction of intact cores from the weakest layers in the rock formation.

SI 22.   The drilling method and sample recovery technique should be designed in relation to the ground to allow optimal assessment of the soil or rock structure. Merely to recover cores of the stronger layers and sludge samples of the weaker layers is unsatisfactory from the point of view of determining the base resistance of piles. A core diameter of 114 mm may be required for very weak rocks such as weathered chalk or marl. One of the principal advantages of obtaining high quality in core recovery is that the cores can be inspected by the piling contractor who can then make his own assessment of pile carrying capacity in a more reliable manner than can be obtained merely by an assessment from descriptions of rock quality in borehole records.

SI 23.   Cores should be adequately preserved and logged; weak rock cores and those subject to disintegration should be photographed on recovery.

BORING THROUGH OBSTRUCTIONS (clause SI 2.02)

SI 24.   The occurrence of boulders or other obstructions in the ground is an important consideration to the practicability of installing any type of pile. If chiselling is successful in breaking through an

---

* Cable percussion.

128

obstruction the time required for chiselling and the weight of the chisel should be stated in the borehole records (clause SI 2.10).

GROUNDWATER OBSERVATIONS (clause SI 2.03)
SI 25.   Information on groundwater levels is vital to the calculation of the carrying capacity of any type of pile in a cohesionless or partly cohesive soil. It is also important in assessing the difficulties of placing concrete in bored and cast in place piles. Some indication of the rate of inflow into a borehole is useful in deciding whether or not it is possible to place concrete in the dry after baling or pumping out a pile borehole. For this purpose simple baling or pouring in tests can be conducted (clause SI 3.073).

SOIL SAMPLING (clause SI 2.04)
SI 26.   The frequency and type of soil sampling required depend on the variation in soil conditions, the number of boreholes and the laboratory testing programme for the particular piling project.

ROCK SAMPLING (clause SI 2.05)
SI 27.   The need for cores of adequate diameter to ensure that complete recovery is obtained is referred to in paragraphs SI 21 and SI 22. The necessity to have a complete core cannot be over-emphasized. Many metres of drilling with no core recovery conveys no worthwhile information. Even partial recovery provides data which are very difficult to interpret. The lost material is presumed to be weaker than that recovered but there is no way of knowing just how weak it is. If correct drilling techniques are employed together with an adequately sized bit it is almost always possible to recover a satisfactory core and the Engineer should not readily be satisfied with less.

GROUNDWATER SAMPLING (clause SI 2.07)
SI 28.   Groundwater samples are required for chemical analysis to ascertain if any substances are present which are deleterious to steel or concrete in piles. It may be desirable to make chemical analyses of samples of soil or rock where salts may be present in crystalline form. It is important that the samples should not be diluted by water added to assist drilling or undergo chemical changes between sampling and analysis in the laboratory.

COMPLETION AND BACKFILLING OF BOREHOLES (clause SI 2.08)
SI 29.   A borehole loosely filled with pervious soil located on a pile position may form a channel for the entry of groundwater to the base of the pile, causing difficulties in obtaining stable conditions and in concreting the pile.

SI 30.   Backfilling of boreholes with bentonite/cement mixes should be undertaken where the presence of an inadequately filled borehole might interfere with a future foundation, particularly where piling may be proposed. Boreholes should always be suitably plugged where they might otherwise penetrate an aquiclude and thus lead to contamination of water in an aquifer or to problems presented by encountering water when engineering works are undertaken within the relatively impermeable ground represented by the aquiclude.

SI 31.   Where boreholes are to be backfilled in areas of present or future underground water supplies advice should be sought from the water authority concerned on the particular measures to be specified to prevent pollution.

TEST PITS (clause SI 2.09)

SI 32.   Test pits are rarely required for piling investigations but they may be needed for close examination of the nature of obstructions in, for example, fill material.

BOREHOLE RECORDS (clause SI 2.10)

SI 33.   Accurate descriptions of soil and rock strata (see paragraphs SI 5 and SI 6) are vital in piling work. Inaccurate or misleading borehole records, more than any other factor, are responsible for claims for extra payment.

IN SITU TESTING

VANE SHEAR TEST (clause SI 3.01)

SI 34.   The vane shear test is essentially a means of measuring the shear strength of very soft and sensitive clays and as such soils contribute little, if anything, to the carrying capacity of piles, its use is of only limited value. However, it may be used either to assess the value of negative skin friction on the shaft of piles embedded in soft alluvial clays which are settling under their own weight or under the weight of superimposed fill, or else to assess the support which laterally loaded piles might obtain.

SI 35.   Vane shear tests may also be of value in assessing the bearing capacity of ground for piling or other heavy plant.

STANDARD PENETRATION TEST (clause SI 3.02)

SI 36.   The standard penetration test should be performed as a matter of routine in all granular soils and also in other materials such as boulder clay, made-up ground, weathered and weak rock from which satisfactory undisturbed samples cannot be obtained. If properly carried out, standard penetration tests can give some indication of pile driving resistance and the results may be correlated with the angle of shearing resistance and relative density of the soil. Hence the shaft friction and end bearing resistance of a pile can be calculated.

SI 37.   In some soils, such as boulder clay containing a high proportion of stones, fill or soil containing cemented layers, from which undisturbed samples cannot be obtained, standard penetration tests are at all times almost the only means available of assessing the consistency, structure and relative density of such materials. They can also give a useful indication of the difference between weathered and relatively unweathered rocks. In unweathered rocks which are sufficiently weak to be penetrated, they can also give an approximate assessment of the relative strength of the material. The test results obtained in weak weathered rocks can be calibrated with a deformation modulus to enable the settlement of piles under working load to be estimated. In the event of standard penetration tests failing to yield reliable results, alternative tests specified by the Engineer may include penetrometer tests (clauses SI 3.03 and SI 3.04).

STATIC CONE PENETRATION TEST (clause SI 3.03)

SI 38.   The static cone penetration test is a useful alternative to the standard penetration test when soil is fine grained or when groundwater conditions make the latter test impracticable or unreliable. If the static cone penetration test is calibrated carefully with a particular type of pile, it can, when used with discretion, give a good guide to the working load of that particular type of pile.

DYNAMIC CONE PENETRATION TEST (clause SI 3.04)

SI 39. Not all penetrometers used in the dynamic cone penetration test are calibrated with the standard penetration test over a wide range of soil conditions and the interpretation of the results is not always easy. A 64 mm cone with a 60° apex angle, driven through 50 mm rods by a 160 kg hammer with a 600 mm drop, gives a penetration resistance in blows/0·3 m of roughly the same order as the standard penetration test in some granular soils.

SI 40. The dynamic cone penetration test is used mainly as a substitute for the standard penetration test when conditions make that test unreliable. It is also very useful when run parallel with standard penetration tests as a check on whether or not the latter are yielding trustworthy results. Like the static cone penetration test it can give unreliable results in soils containing occasional stones or fragments of rock.

SI 41. Although the penetrometer can penetrate weathered and weak rocks more readily than the static cone, it is not normally used for this purpose.

USE OF PENETRATION TESTS BY STATIC OR DYNAMIC METHODS

SI 42. Penetration tests can give unreliable results in soils containing occasional stones or fragments of rock; their best use is in the prediction of the driving resistance, skin friction and bearing resistance of driven piles in cohesionless soils. They are less useful for bored piles where the effects of pile installation cannot be reproduced by the test. They are used for rocks or weathered rocks only in which a measurable penetration can be attained.

PRESSUREMETER TEST (clause SI 3.05)

SI 43. The pressuremeter consists essentially of a cylindrical membrane which is expanded to compress the walls of a borehole. The deformation of the soil or rock is measured by observing the increase in volume of liquid injected into the membrane or by monitoring the increase in the diameter of the cell. The ultimate bearing capacity and a deformation modulus of the soil or rock are obtained from the observed volume increase.

SI 44. The use of the pressuremeter is limited in the design of piled foundations. Where clay soils are concerned traditional sampling and laboratory testing, of which there is considerable practical experience, are to be preferred. The pressuremeter is most useful in a soil such as boulder clay which contains a high proportion of stone or in weathered rock, both of which are unsuited to most other forms of test. As such the pressuremeter test is a possible alternative to plate bearing tests at the bottom of a borehole as a means of assessing the end bearing resistance of a pile, although it is of little value in estimating shaft friction. It is very useful in assessing the resistance of piles to lateral loads.

SI 45. Development of the pressuremeter may permit separate measurement of total pressure and pore pressure. The Camkometer can be inserted into the soil with little disturbance. It is a promising tool for the assessment of conditions in offshore investigations. In view of its limitations, guidance on the suitability of ground for investigations with a pressuremeter should be sought from a specialist contractor, unless the Engineer has specialist experience. The responsibility for the interpretation of the results should be settled by agreement and clearly stated in writing before the work is ordered.

BOREHOLE PLATE LOADING TEST (clause SI 3.06)

SI 46. The vertical borehole plate loading test is expensive but it is by far the most reliable in soils such as gravels, boulders or rocks which are unsuited to standard laboratory tests.

SI 47. For end bearing piles on rock where it may be possible to consider mobilization of a high fraction of the compressive strength of the concrete in the pile, a plate bearing test may be the only practical means of confirmation.

SI 48. This test may also be used to determine settlement of an end bearing pile on rock. Where boreholes or trials pits are large enough to enable the plate to be carefully bedded to the surface, the load/settlement curves will provide information on the deformation modulus of the ground. The settlement of piles over a range of sizes can then be calculated.

SI 49. Where tests are made in small boreholes in the manner specified, it is doubtful whether the results will be of sufficient accuracy to establish a reliable deformation modulus. In such cases the tests should be taken to failure load and the allowable end bearing pressure for the pile established by applying a suitable factor of safety to the measured ultimate bearing resistance.

SI 50. The results of tests to determine shaft friction are applicable only to bored piles where the boring and concreting for the test are undertaken in a manner similar to that proposed for the piles. The time interval between boring and concreting for the shaft friction tests should be of the same order as that proposed for the piles. The shaft friction tests as specified cannot be used to estimate allowable values for any form of driven piles. The shaft must be checked to see that it is safe for men to enter; reference should be made to CP 2011.

VARIABLE HEAD PERMEABILITY TEST (clause SI 3.073)

SI 51. The variable head permeability test undertaken by baling or pouring in methods is used mainly in connection with the construction of bored cast in place piles. It gives valuable guidance on the quantity of water likely to flow into the borehole, thus indicating whether or not it will be possible to cast the pile in the dry. The baling or pumping out method is to be preferred, as the data it gives are more directly applicable than those which can be obtained from the pouring in method. However, the latter method is cheaper and quicker; it gives only a qualitative result but even this is better than speculation.

SI 52. In situ permeability tests are also of value in estimating the rate of settlement of large pile groups founded in compressible soils. Because of stratifications and laminations in some compressible soils, the horizontal permeability is much greater than the vertical permeability. Consolidation tests made by vertical loading on small specimens do not always give a true measure of the in situ permeability of such deposits; hence the rate of consolidation of deep foundations as calculated from these tests will be slower than the actual rate.

GEOPHYSICAL METHODS OF SITE INVESTIGATION
(clauses SI 3.08–SI 3.11)

SI 53. Geophysical methods of site investigation provide a wide choice of techniques. For surveys over wide areas, gravity surveys and magnetic surveys may reveal major hidden structural features of the ground. There are also numerous logging techniques that may be used in boreholes for specific purposes. The geophysical tests most likely to be considered for a site investigation for a foundation use seismic or resistivity techniques. Each type will provide results which can be interpreted usefully only for particular conditions of the ground, especially where the ground is layered in a relatively simple manner and where the appropriate physical properties of the layers are distinctly different. For resistivity and seismic surveys it is necessary that the electrical resistivity or seismic velocity should increase with depth. It is normally imperative that the results from geophysical methods should be calibrated against direct information obtained from boreholes. Where conditions are favourable and large areas are to be covered by an investigation, the cost of a geophysical survey may be much less than the additional cost of sinking further boreholes to provide comparably detailed information.

SI 54. There are many forms of seismic and resistivity survey, varying greatly in cost and in detail of information provided. The choice and type should be carefully related to the object of the specific investigation. The electrical resistivity method may be used to relate corrosion of buried structures to the resistivity of the ground. Continuous reflection seismic profiling is a valuable technique in assessing foundation conditions below water, for which, in favourable conditions, interfaces between different strata may be continuously recorded. The particular type of apparatus and the manner in which it is used should be related to the depth of water, depth of penetration required and nature of the ground; these are matters on which specialist advice should be sought.

LABORATORY TESTING
REPORTING TEST RESULTS (clause SI 4.02)

SI 55.   Sieve and sedimentation analyses are rarely required in connection with piling investigations. They are needed if some special ground treatment processes (e.g. cement or chemical injections) are involved in the piling technique.

SI 56.   If part of a scheme includes dredging, sieve and sedimentation analyses, standard penetration test results and in situ densities are required.

SI 57.   Shear strength determinations obtained by undrained triaxial or unconfined compression tests are required to calculate the skin friction and end bearing resistance of piles in cohesive soils. Drained tests to determine values of effective cohesion and shearing resistance may be required for special investigations of the stability of individual piles or pile groups but the full significance of these parameters in respect of the carrying capacity of piles is not yet understood. Consolidation tests may be needed to calculate the settlement of groups of piles in cohesive soils.

SI 58.   Unconfined or triaxial compression tests on rock cores are of little value in determining the carrying capacity of piles but they may be of use to the piling contractor in assessing difficulties in drilling through rock, or in assessing the likely refusal level of piles driven into a rock stratum.

SI 59.   Chemical analyses are required to determine the sulphur trioxide content, the pH value and whether or not any special precautions are needed against attack by aggressive substances on concrete or steel. Full chemical analyses and possibly bacteriological analyses may be needed in connection with corrosion studies or in cases of potential attack by chemical wastes on piles in fill material.

MODEL SPECIFICATION FOR SITE INVESTIGATION FOR PILING

SECTION SI 1. PARTICULAR CLAUSES

SI 1.01. GENERAL

The following information is pertinent and particular to this investigation

(*a*) location of the site (* *the situation of the Site and available relevant geological information and reference thereto*)

(*b*) existing conditions on the Site (* *the surface and subsurface conditions, existing and previous structures, factors affecting groundwater levels and nearby structures which the investigation might disturb and any other relevant conditions*)

(*c*) proposed development (* *the nature of the intended final development of the Site*)

(*d*) access to the Site (* *facilities, requirements and limitations relating to access and occupation of the Site by the Contractor*)

(*e*) setting out (* *the datum used and points for setting out*)

(*f*) obstructions and prevention of damage to services (* *details of known or suspected services above or below ground as referred to in clauses SI 1.04 and SI 2.02*)

(*g*) marine borings*

(*h*) wharfage and berthing facilities*

SI 1.02. REQUIREMENTS

Particular requirements of this specification are

(*a*) type of investigation and report required (* *the extent of the investigation and the form of the report required*)

(*b*) boreholes (* *the number, description and location of boreholes together with the required depths and sampling depths*)

(*c*) in situ tests (* *the number, description and location of all in situ tests required*)

(*d*) laboratory tests (* *the types of tests required and the scope of the testing work*)

(*e*) responsibility—the Contractor is responsible for the accuracy in setting out to the data provided, for the accuracy of depth records to given bench marks on the Site, for the accuracy of his observations generally and for his reports on the observations

SI 1.03. SUPERVISION

The Contractor's Agent for the work shall be, or alternatively he shall have on his staff, a soils engineer, engineering geologist, engineer or engineering technician fully experienced in site investigation work.

---

* The Specification writer should insert particular information here; words in italics describe the information to be provided.

This specialist shall make frequent visits to the Site and at other times be readily available by telephone to both the drillers and the Engineer. Where stated in the Particular Specification or where directed by the Engineer, he shall be on the Site full time.

SI 1.04. SERVICES AND STRUCTURES

The Contractor shall take all reasonable precautions to avoid damage to services above and below ground, such as sewers, drains, gas and water mains, electricity and other cables and all structures above and below ground. He shall be responsible for any damage caused to services, the location and height or depth of which have been notified to him.

The Contractor shall ascertain from the local authority and statutory undertakers the information available concerning services and structures before commencing work on the Site. Where underground services or structures are thought to exist in the immediate vicinity of the position of a borehole, a test pit shall be dug and taken to such a depth that it is reasonably certain that all underground services and structures have been uncovered. After this normal boring may proceed.

SI 1.05. POLLUTION

The Contractor shall not foul the atmosphere, any river, stream, watercourse or sewer or contravene section 2 of the Rivers (Prevention of Pollution) Act 1951. He shall make provision for the discharge or disposal from the Works or Temporary Works of all water waste products and spoil however arising. The method of disposal shall be to the satisfaction of the Engineer and of any other authority or person having an interest in any land or watercourse over or in which water and waste products may be discharged.

The requirements of this clause shall not limit any of the Contractor's statutory obligations or liabilities.

SI 1.06. POSITION OF BOREHOLES

The position of boreholes and standpipes as made shall be fixed from features of the Site for ease of reference and their co-ordinates shall be recorded in the report on the investigation.

SI 1.07. OBTAINING AND STORING SAMPLES

Methods of sampling and the subsequent handling, labelling and storage of samples shall be in accordance with the procedure described in CP 2001/57.

SECTION SI 2. BORING AND SAMPLING

SI 2.01. BORING METHODS

All boreholes shall be supported with casing as may be necessary to prevent collapse of the sides. The Contractor shall carry out the work in such a way that the groundwater levels over the Site are not generally affected and shall take all necessary precautions to prevent surface water entering the boreholes.

Boring in soils and in rocks where specified shall be to a minimum diameter of 150 mm and shall be undertaken either by shell and auger (cable percussion) or by a rotary power auger of a type which enables samples of at least 100 mm dia. to be obtained using open or piston samplers as specified in CP 2001. Drilling in rocks shall be undertaken by rotary coring methods to obtain cores of not less than the minimum diameter stated in the Particular Specification.

SI 2.02. BORING THROUGH OBSTRUCTIONS

If an obstruction in the form of bedrock, boulder, concrete, brickwork, timber or other natural or man-made object is encountered which prevents further progress in boring by shell and auger or rotary power auger, the Contractor shall attempt to break through the obstruction by chiselling. If the size or composition of the obstruction is such that little or no progress is made by chiselling, the Contractor shall consult with the Engineer who may instruct him to use rotary coring methods to drill through and obtain cores of the obstruction, in which case the cores shall have a diameter of not less than that stated in the Particular Specification. If the boring shows that the obstruction is bedrock, the rotary core drilling shall be continued to the depth required by the Engineer and to the diameter specified to prove the continuity and engineering characteristics of the formation.

If the obstruction is shown to be a boulder, ledge of rock or other object underlain by soil, the Contractor shall consult and agree with the Engineer the use of one of the following lines of action.

(*a*) Chisel out the cored borehole through the obstruction sufficient to allow shell and auger boring, in situ sampling and testing to continue below the obstruction.

(*b*) Continue the boring by rotary core drilling to the required depth of the borehole at the diameters referred to in the Particular Specification. Then the Contractor shall consult with the Engineer as to whether or not it is necessary to obtain undisturbed samples of the soils in a nearby borehole at the levels beneath the obstruction.

(*c*) Abandon the borehole and drill another one near by to obtain the necessary samples.

140

SI 2.03.   GROUNDWATER OBSERVATIONS

When groundwater is encountered in a borehole, the Contractor shall immediately cease boring and observe and record any movement in the level of the groundwater. Boring shall cease for half an hour, at the end of which period the water level shall again be recorded. The Engineer's instruction shall then be sought and, if required by him, a standpipe or piezometer shall be installed on completion of the borehole (as detailed in this clause).

Changes in water level in any deeper aquifers shall be observed and recorded in a manner similar to that already specified. Records of groundwater levels shall be made as specified in clause SI 2.10, and in all cases shall be referred to the datum used for the Works.

If casing is used and this forms a seal against the entry of groundwater, the Contractor shall record the depth at which no further entry or only insignificant infiltration of water occurs.

If it is necessary to add water to the borehole to facilitate boring this shall be used sparingly and not in such a manner as to prevent accurate observation of the groundwater conditions in the borehole.

The Contractor shall install standpipes in boreholes to the numbers and depths required by the Engineer. A standpipe shall comprise a rigid polyvinyl chloride or galvanized iron pipe of 25 mm diameter having at its lower end an approved porous filter element. The filter shall be surrounded by sand or sand and gravel mixture graded in such a manner as to prevent entry of soil particles into the filter. A layer of the graded mixture should be placed in the bottom of the borehole prior to lowering the standpipe. The sand or sand and gravel mixture shall be placed in position before withdrawing the borehole casing.

Where instructed by the Engineer, the Contractor shall fill the annulus above the filter with a 0·5 m deep layer of bentonite–cement grout. A plug of stiff bentonite, in the form of balls dropped down the borehole, shall be carefully tamped into place, followed by pumping down of a bentonite–cement grout to fill the borehole to a level of at least 0·5 m above the filter element. The Engineer may also instruct the Contractor to form a plug of bentonite–cement mix of stiff plastic consistency to be carefully tamped into place below the level of the standpipe to form a seal against upward flow of groundwater into the standpipe from a deeper aquifer. The top of the standpipe shall be provided with a screwed cap, which shall be not less than 150 mm below ground level. The cap shall be protected by a short length of 150 mm pipe set vertically with its upper end covered by a paving slab set flush with the ground surface.

During the period he is on the Site the Contractor shall take daily readings of water levels in all standpipes or, if so directed, at more frequent intervals. He shall also, where directed by the Engineer and as required by the Specification, return to the Site at intervals to make additional water level readings.

SI 2.04. SOIL SAMPLING

Undisturbed samples of cohesive soils shall be obtained at each change of stratum and at a spacing of not more than 1·2 m in the boreholes unless otherwise instructed by the Engineer.

A standard penetration test, as specified in clause SI 3.02, shall be made to recover a disturbed sample of cohesionless soil on encountering a stratum of such soil and at intervals of not more than 1 m in the stratum unless otherwise instructed by the Engineer.

Where undisturbed soil samples and standard penetration test samples are not obtained the Contractor shall recover disturbed samples of soil from the boring tools. The disturbed samples shall be obtained at such spacing as to ensure that samples from the borehole either in the form of undisturbed samples, standard penetration test samples or disturbed samples are obtained for every 1 m depth bored. The minimum weight of disturbed samples shall be as specified in BS 1377, clause 1.5.3.

SI 2.05. ROCK SAMPLING
SI 2.051. Shell and auger method

Where boreholes are bored into rock using shell and auger or rotary power auger equipment, samples of rock in the form of cuttings or sludge shall be obtained on encountering the rock stratum, at each change of rock stratum and at intervals of not less than 1 m in the boreholes. The samples shall weigh not less than 0·5 kg each and shall be labelled and stored in accordance with the procedure described in CP 2001/57, appendix G.

SI 2.052. Rotary coring methods

Where boreholes are bored using rotary coring methods, rock cores of diameter not less than the diameter stated in the Particular Specification shall be extracted. After being brought to the surface they shall be removed from the core barrel by methods designed to cause the least possible further disturbance. Where split inner core barrels are not in use the core shall be extracted by means of steadily applied pressure. Extraction by means of hammering the barrel or explosive extrusion under high air pressure or water pressures will not be permitted. After extrusion the core shall be placed in a purpose-made core box.

SI 2.053. Core boxes

Core boxes shall be of sound, robust construction and shall be able to withstand the weight of core and any full boxes which may subsequently be placed on them, and sufficiently watertight to protect the core from rain. They shall be specially made to hold the size of core which is being obtained tightly in place in rows separated by wooden slats. The boxes shall be provided with a strong metal hinged lid fitted with a padlock, hasp and staple for closing and shall have end ropes for handling. The top and bottom of the boxes shall be reinforced by cross straps to aid stacking and retrieval. They shall be constructed of wood, marine plywood or other material approved by the Engineer.

SI 2.054. *Placing cores in boxes*

As cores are extruded they shall be laid in a core box with the shallowest core to the left and the deepest to the right. The highest row of core shall be placed near the lid hinge.

The box shall be identified inside and outside by the Site name, borehole number, core box number, depth of top and bottom of core included, the Contractor's name and the date. This information shall either be painted on the box or stamped on metal labels, which shall be waxed and nailed to the box. The depth of the top and bottom of the total core and separate core runs shall be painted on blocks of wood or other material approved by the Engineer made to fit between the dividing slats.

In order that zones of core loss can be readily identified, wooden dowels cut to appropriate lengths and suitably identified shall be inserted either in the sections where loss occurred or at the base of the particular core runs.

Sections of core which are weak and friable, formed of rocks, or soils which are likely to dry out or otherwise alter in nature with time, shall be first sealed with aluminium foil and subsequently covered with wax before being placed in the core box.

SI 2.06. *HANDLING SOIL AND ROCK SAMPLES*

Care shall be taken in handling and transporting the samples of soil and cores of rock from the Site to the Contractor's or another testing laboratory. This shall, where possible, be done using the Contractor's transport manned by skilled personnel and the Contractor shall arrange that whoever is responsible for transporting samples is informed of the need to avoid unnecessary disturbance.

SI 2.07. *GROUNDWATER SAMPLING*

A sample of groundwater shall be taken as soon as sufficient water has entered the borehole after boring has reached groundwater level. If water has been added to the borehole before reaching groundwater level, all water in the borehole shall be baled out and the uncontaminated groundwater shall be allowed to seep back into the borehole before a sample is taken.

Where the groundwater is sealed off by the borehole casing and a lower aquifer is encountered a sample of water from this and any succeeding aquifers shall be taken as already specified.

Samples of groundwater shall be at least 500 ml in volume and shall be placed in clean jars or bottles which shall be labelled and stored as described in CP 2001/57.

SI 2.08. *COMPLETION AND BACKFILLING OF BOREHOLES*

The completion of boreholes and the start of backfilling shall be approved by the Engineer unless he has given prior instructions on the depth at which the boreholes are to be terminated.

The soil backfilled into the boreholes shall be placed in layers, each successive layer being tamped by the boring tools.

Where instructed by the Engineer, the boreholes shall be backfilled with concrete having a cement content of not less than 250 kg/m^3.

Backfilling shall be done as the casing is withdrawn.

SI 2.09. *TEST PITS*

Test pits having a plan area of not less than 1·5 m^2 shall be excavated in positions as instructed by the Engineer. The Contractor shall provide all pumping and take all measures (e.g. sheet piling and timbering) to ensure stability of adjacent structures or installations and safe conditions for descent into and working in the pits. While observations are in progress the Contractor shall take all necessary steps to exclude water from the test pits.

SI 2.10. *BOREHOLE RECORDS*

The Contractor shall supply to the Engineer not later than noon on the following working day a journal of each day's boring. Where boring is by shell and auger or rotary power auger the journal shall state

(*a*) date and times of boring

(*b*) borehole number

(*c*) ground level at position of boring

(*d*) type of plant used and method of boring

(*e*) diameter of boring

(*f*) diameter and depths of casing

(*g*) all water levels encountered, including measurements of fluctuation of adjacent tidal waters, lakes or rivers

(*h*) depths at which groundwater was sealed off

(*i*) level at which groundwater was standing at commencement and termination of working day (where a boring is on land but is near tidal waters the level of which varies the level of those waters is to be recorded at intervals throughout the day)

(*j*) levels at which water, if any, was added to borehole

(*k*) levels of water in standpipes

(*l*) depths of base of each stratum from ground level and a preliminary description of the strata

(*m*) depths at which samples of all types were taken

(*n*) records of in situ tests made and instrumentation installed

(*o*) time for chiselling through obstructions and weight of chisel

(*p*) other relevant remarks.

Where boring is by rotary coring methods, the journal shall state

(*a*) date and times of boring

(*b*) borehole number

(*c*) ground level at position of boring

(*d*) type of plant used and method of operation, including details of type of flushing

(*e*) type of core barrel and bit

(*f*) depth of hole at start and end of working day or shift as relevant

(*g*) depth of start and finish of each core run

(*h*) depth and size of casing at start and end of each core run

(*i*) core diameter and changes in core size

(*j*) state of bit

(*k*) time to drill each core run

(*l*) character and proportion of each flush return

(*m*) level at which groundwater was standing at commencement and termination of working day (where a boring is on land but is near tidal waters or near waters the level of which varies the level of those waters is to be recorded at intervals throughout the day)

(*n*) depths of base of each stratum from ground level and a preliminary description of the strata

(*o*) total core recovery with information as to possible location of core loss

(*p*) details of in situ tests and instrumentation installed

(*q*) other relevant remarks.

Where boring is from marine craft other relevant remarks shall include

(*a*) time and height of high and low water at a tide gauge and tidal heights at intervals as directed by the Engineer

(*b*) time and height of water levels at borehole position at intervals as directed by the Engineer

(*c*) details of movement of water levels within borehole related to fluctuation of water levels at borehole location

(*d*) detailed records of delays due to reasons other than failure of boring equipment (e.g. craft dragging anchors, mist, shipping movements and broken drilling tubes)

145

(*e*) detailed records of movement, vibration and oscillation of drilling tubes

(*f*) detailed records of bowing of drilling tubes because of waves, tidal currents and so on.

SI 2.11. *FINAL BOREHOLE LOGS*

After completion of all soil tests and a visual examination of all the samples in the laboratory, final borehole logs shall be prepared which shall include a grid or topographical reference and details similar to those in clause SI 2.10 but corrected in the light of all the information finally available, and in the light of description by a qualified soils engineer or engineering geologist. The soil descriptions shall conform to CP 2001/57 and rock descriptions shall conform to section 5 of the report 'The logging of rock cores for engineering purposes'.*

The borehole logs shall include a grid or topographical reference of the borehole.

In the case of cores obtained by rotary drilling methods the final borehole logs shall state in addition to the information given in clause SI 2.10 corrected as necessary

(*a*) condition of each core run in terms of percentage recovery and rock quality designation, i.e. the percentage of solid core recovered in pieces longer than 100 mm

(*b*) definition of the rock type, its alteration state and relative strength; details of the natural discontinuities and rock structures.

---

* Geological Society Engineering Group Working Party Report. *Q. J. Engng Geol.*, 1970, **3**, No. 1, 1–24.

SECTION SI 3. IN SITU TESTING

SI 3.01. VANE SHEAR TEST

The vane shear test shall be carried out in accordance with BS 1377, test 18. After completion of the standard test, if specified, the remoulded shear strength of the soil shall be measured. The torque measuring instrument shall be disconnected from the extension rods and the vane turned through at least six complete revolutions. A period of at least five minutes shall elapse and then the vane test shall be repeated.

SI 3.02. STANDARD PENETRATION TEST

The standard penetration test shall be carried out in accordance with BS 1377, test 19. If, in spite of all the precautions specified to avoid carrying out the test in loosened soil, there is good reason to believe that unrealistically low results are being obtained or soil flows into the borehole so fast as to prevent the test being carried out, the Contractor shall report the matter to the Engineer and seek his instruction as to whether or not an alternative test is to be undertaken.

SI 3.03. STATIC (DUTCH) CONE PENETRATION TEST

SI 3.031. Apparatus

The penetrometer for the static cone penetration test shall consist of a rod with a cone and a sleeve which surrounds the rod so arranged that both cone and sleeve can be advanced together or independently of each other. Alternatively the cone shall be fixed to the end of a hollow rod without means of independent movement. The cone shall have an apex angle of 60° and a cross-sectional area of $0\cdot01$ m². A mantle shall be provided above the cone.

The penetrometer shall be forced into the soil by means of a static load. Where the required depth of penetration is not great and the soil is suitable, the static load may be provided manually. Elsewhere the load shall be applied by mechanical means using as resistance either kentledge, screw anchors or other approved means.

SI 3.032. Operation of apparatus

The load applied to the penetrometer shall be measured by a pressure gauge, strain gauges or other means approved by the Engineer.

To carry out the test first the cone and sleeve of the penetrometer shall be driven together into the soil to measure the combined cone and shaft resistance. The cone shall then be advanced independently of the sleeve to measure the cone resistance alone. Tests shall be carried out at convenient intervals, which shall not generally exceed 250 mm, until the required depth of penetration has been reached.

Alternatively, the hollow rod fixed to the cone shall be pushed down in continuous movement. In this case the resistance of the soil to thrust by the cone and the skin friction on the mantle above the cone shall be measured continuously by means of electrical resistance strain gauges installed within these components.

If progress becomes impossible with either apparatus before the specified depth has been reached, a borehole shall be drilled to the level where progress ceased and the test shall be continued from the bottom of this borehole or from successive boreholes drilled in like manner until the specified depth has been reached.

If the Contractor intends to depart from this method of testing he shall state this in his tender.

SI 3.033. *Results of tests*

The results of each test shall be presented in the form of three graphs

(*a*) combined penetration resistance of cone and sleeve plotted against depth

(*b*) penetration resistance of cone plotted against depth

(*c*) penetration resistance of sleeve (difference between (*a*) and (*b*)) plotted against depth.

If specified, the Contractor shall provide an interpretation of the results.

SI 3.04. *DYNAMIC CONE PENETRATION TEST*

SI 3.041. *Apparatus*

The dynamic cone penetration test shall be carried out independently of any borehole. The apparatus shall comprise a sectional rod with a cone fitted at the end, the base of the cone being of slightly greater diameter than that of the rod. It shall be driven into the ground by a constant weight dropped through a constant height. The apparatus shall be purpose-made and the weight shall always fall automatically through a constant height.

The Contractor shall give, at the time of tender, a full description of the apparatus he intends to use, its mode of operation and its calibration with the standard penetration test.

SI 3.042. *Results of tests*

The results of the dynamic cone penetration test shall be presented in the form of a graph of depth plotted against number of blows or any convenient unit of penetration.

SI 3.05. *PRESSUREMETER TEST*

The Contractor or Sub-contractor shall ensure that pressuremeter tests are carried out by those competent in the testing technique and its interpretation. The tests shall be made in boreholes as directed by the Engineer after the results of preliminary boreholes are available. The number of tests and the depths at which they shall be made shall be the subject of discussion and agreement between the Engineer and the Contractor and/or the Sub-contractor.

The Sub-contractor shall give to the Contractor for the Engineer an interpretation of the test data in the form of ultimate bearing capacity and settlement under working load of piles of specified diameter.

SI 3.06. *BOREHOLE PLATE LOADING TEST*

SI 3.061. *Position of tests*

Borehole vertical plate loading tests shall be carried out at the depths or on the strata stated in the Specification. The borehole may be sunk by any plant suited to its size and its diameter shall be as stated in the Specification.

SI 3.062. *Apparatus*

A circular plate of diameter nearly equal to that of the borehole shall be used for the plate loading test. For a borehole of diameter 600 mm or less the diameter of the plate shall not exceed the diameter of the borehole less 25 mm. For a borehole of diameter greater than 600 mm the diameter of the plate shall not exceed the diameter of the borehole less 50 mm.

The plate shall be loaded through a column formed by a steel tube or hollow pile co-axial with the plate. The load shall be applied to the column by means of a hydraulic jack operating against kentledge co-axial with the column or two or more tension piles or ground anchors. The load on the plate shall be measured by an approved load cell or pressure capsule capable of being read to an accuracy of 1%. Before and after use the load cell or pressure capsule shall be calibrated over the expected range and actual range of loading by an approved testing laboratory. A copy of the calibration certificates shall be supplied to the Engineer.

The movement of the plate under load shall be transmitted to dial gauges at the surface by means of a settlement measuring rod located within the steel tube by which the load is applied and which shall have observation slots.

The settlement measuring rod shall be restrained from lateral movement by means of rod guides fixed inside the steel tube. The dial gauge shall record to 0·02 mm and shall be arranged in contact with a small disc of machined steel attached to the settlement measuring rod. The dial gauges themselves shall be attached to a reference beam supported on two foundations sufficiently far from the borehole not to be influenced by any movement of the casing.

SI 3.063. *Resistance to test load*

Where kentledge is used it shall be supported on a properly designed frame such that the load cannot tilt or collapse.

Where tension piles or ground anchors are used, they shall be sited sufficiently far from the centre of the borehole not to influence the behaviour of the plate or surrounding ground but the centres of the piles or anchors shall be not closer to the centre of the borehole than three times the diameter of the plate.

The jacking resistance provided shall be sufficient to allow a pressure of at least three times the expected allowable end bearing capacity to be applied in the case of soils or weathered rock.

In the case of an intact or unweathered rock, or rock containing few joints, the pressure applied to the rock shall be a maximum of $10 \, MN/m^2$.

SI 3.064. *Termination of casing*

The plate loading test shall be made with the borehole casing extended to the full depth of the borehole, although the Engineer may permit the casing to be terminated above this level if the groundwater has either been sealed off or if there is none present and if the material below the casing is sufficiently stable to stand on its own without any fragments falling into the bore.

SI 3.065. *Water level*

Where it proves impracticable to seal off the groundwater in the borehole, the water level may be temporarily lowered by pumping below the specified level of the plate to enable it to be installed in dry conditions. The water shall then be allowed to return to its normal rest level before the test is started. Pumping of water from the borehole or pumping from adjacent wells shall not be permitted if the resulting groundwater flow causes changes in the density or other characteristics of the soil or rock beneath the test plate.

Sumps formed for lowering the groundwater level below the proposed level of the plate shall be such that the characteristics of the ground are unaffected by their use.

SI 3.066. *Setting of plate*

The bottom of the borehole shall be thoroughly cleaned to remove loose or disturbed soil or rock before the plate is bedded. This shall be done manually if the size of the borehole enables a man to be lowered to the bottom. After manual cleaning of the borehole, a 25 mm minimum thick layer of neat cement mortar of stiff consistency shall be rammed into place to cover the area to be loaded by the test plate and shall then be carefully trowelled to a level surface. After the mortar has set, a thin layer of plaster of Paris shall be poured on and allowed to set. Finally a second thin layer of plaster of Paris shall be poured on. The plate shall be bedded on to the second layer before it has set.

If the diameter of the borehole is too small to permit manual cleaning and bedding of the plate, the cleaning shall be carried out by an approved auger or hinged bucket operated at the end of a drill rod assembly. After completion of cleaning in this manner, a layer of neat cement mortar shall be placed in the bottom of the borehole by a bottom-opening bucket or the pipe of a tremie. Before the mortar has set the plate shall be lowered into the hole and lightly pressed on to the surface of the mortar to bed the plate.

SI 3.067. *Operation of apparatus*

The load shall be applied in five equal increments up to and including the maximum expected load, and then released. Each increment of load shall be maintained until the settlement of the plate has ceased or has slowed to a rate not exceeding 0·1 mm/h or until continual settlement denoting failure in shear of the soil or rock beneath the plate has occurred. The temperature records of the measuring rod shall be taken at the time of each series of settlement observations.

Where, in addition to data on the allowable end bearing capacity of piles, an assessment of shaft friction is required, the plate shall be removed from the bottom of the borehole and replaced either by a layer of compressible material or by a suitably designed collapsible container. The shaft shall be filled with concrete as the casing is withdrawn. The load shall then be applied and the settlement measured as already described.

Where the shaft friction of only part of the soil profile is required to be determined, as in the case of rock sockets, the concrete shall first be brought up to the level of the top of the stratum concerned. After this a permanent steel casing of lesser diameter than that in contact with the soil shall be introduced into the bore and the remainder of the pile cast within this casing so forming a concrete column. The loading test shall be carried out as described at least seven days after the test section is concreted. On completion of the plate loading test the Engineer shall instruct the Contractor on measures to be taken to fill the space between the column and the sides of the borehole.

SI 3.068. *Results of tests*

The results of plate loading tests shall be supplied in the form of graphs as follows

    (*a*) *plate bearing tests*: pressure under plate (kN/m^2) against settlement (mm) and time of application of load against settlement (mm)

    (*b*) *shaft friction tests*: shaft friction (kN/m^2) plotted against settlement (mm).

SI 3.07.   *IN SITU PERMEABILITY TEST*

SI 3.071. *Constant head test*

The constant head test shall be carried out as directed by the Engineer during the sinking of boreholes. The general arrangement for the test is shown in Fig. 1.

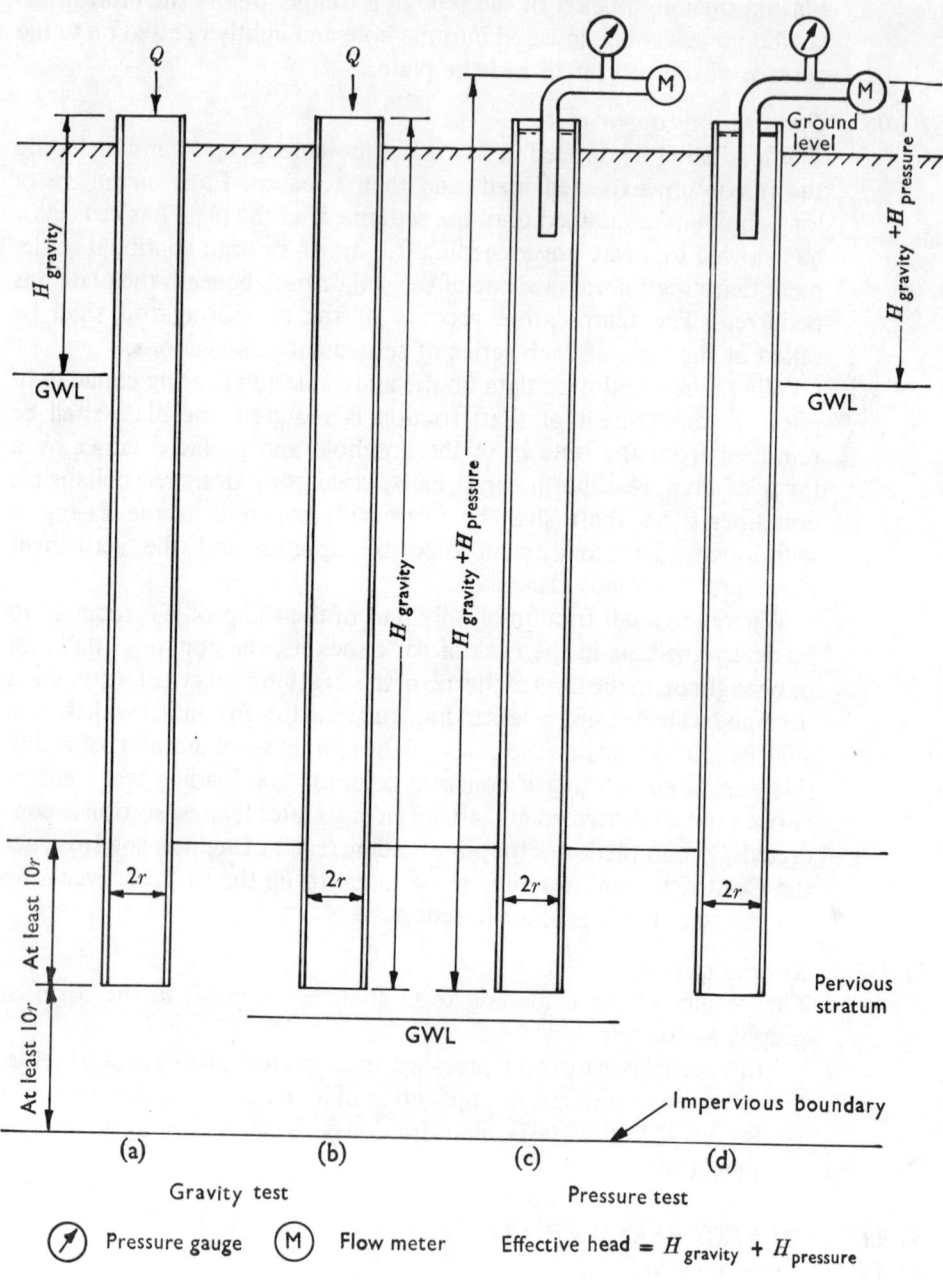

Fig. 1. Open end pipe test for soil permeability

After reaching the level of a pervious stratum in which a test is to be made, boring shall continue until the penetration into the stratum is not less than five times the internal diameter of the casing being used. The borehole shall then be cleaned out just to the bottom of the casing. If the casing is below the level of the groundwater, it shall be kept full of water during cleaning and withdrawal of tools so that soil does not squeeze into it.

A metered supply of fresh clean water shall be discharged into the borehole. This water shall be free from silt or clay particles to avoid the possibility of the interstices of the test section becoming plugged. If necessary, the water shall be freed of foreign matter either by means of filtration or by the use of settlement tanks.

Where practicable, the temperature of the added water shall be higher than that of the groundwater so as to preclude the creation of air bubbles in the ground which might greatly reduce the acceptance of water.

In gravity tests, the rate of flow of the water supplied to the borehole shall be adjusted so that the water level remains constant against a mark inside the borehole casing near the top. If it is impossible to maintain the water level a constant rate of flow producing a fluctuation of no more than 75 mm above and below the mark during a period of 5 minutes will be accepted. When the steady state has been reached, the quantity of water entering the borehole over a fixed interval of not less than 5 minutes shall be recorded.

SI 3.072. *Pressure test*

Where directed by the Engineer, a pressure test shall be carried out. The general arrangement of the test is shown in Fig. 1(c) and (d). A suitable pump shall be introduced into the circuit together with a pressure gauge, and the top of the borehole casing adequately sealed. The test shall then be undertaken in the manner specified in clause SI 3.071.

The following records shall be kept.

(a) Q: rate of flow of water into the borehole (m^3/s).

(b) r: internal radius of borehole casing (m).

(c) H: head (m) causing flow. For the gravity test, this is the distance indicated in Fig. 1(a) and (b). For the pressure test the head is the sum of the gravity head as shown in Fig. 1(c) and (d) plus the pressure recorded by the pressure gauge converted into head of water. The permeability of the strata (m/s) is then given by $K = Q/5 \cdot 5rH$.

SI 3.073. *Variable head test*

Where directed by the Engineer the variable head test shall be carried out in place of the constant head test.

Where conditions in the borehole indicate that a permeable stratum has been reached, the borehole shall be baled or pumped out in order to lower the level of the water which shall then be recorded at intervals until it returns to equilibrium.

If it is impossible to lower the water level in the borehole, either because the soil is too permeable or because it flows into the casing, a pouring in test shall be carried out instead. In this test water shall be poured into the borehole until the casing is full. The water level shall then be recorded at intervals until the equilibrium position is reached.

The results of these tests shall be presented in the form of a simple table showing water level and time. In each case the groundwater level before and during the test shall be recorded.

Geophysical methods of site investigation

SI 3.08. *GENERAL*

When geophysical methods are specified the Contractor shall refer to the Particular Specification for information available to the Engineer about the geology of the Site and data of the soil profile. The Particular Specification shall state the purpose of the investigation and the information it is expected to yield. The geophysical method will be specified but the Contractor shall state in his tender whether he considers it to be the most appropriate or whether he proposes to use another method.

The Contractor shall provide an interpretation of the results of the investigation.

Clauses SI 3.09, SI 3.10 and SI 3.11 describe apparatus in general use and the way in which the tests are normally carried out. However, as much of the apparatus is proprietary, another method proposed by the Contractor would not necessarily be rejected if it did not comply precisely with every appropriate clause, but where differences do occur, these shall be clearly stated at the time of tender.

SI 3.09. *ELECTRICAL RESISTIVITY METHOD*

SI 3.091. *Apparatus*

The essential components of the apparatus for the electrical resistivity test shall be

(*a*) four metal electrodes capable of being driven into the soil

(*b*) a source of low frequency alternating current with cables to connect it to each of two of the electrodes

(*c*) a potentiometer or other electrical resistance measuring instrument also with cables to connect it to each of the other two electrodes.

154

The Contractor shall state at the time of tender which type of apparatus he intends to use. He shall also give a full description of it, together with a statement of his experience of it for the type of work required by the Particular Specification.

SI 3.092. *Method of carrying out test*
The methods of arrangement of the electrodes shall be as specified in clause SI 3.093 and the two basic techniques of carrying out the tests shall be as specified in clauses SI 3.094 and SI 3.095. The exact procedure adopted shall be that best suited to produce the required data. If the exact procedure is not laid down in the Particular Specification, the Contractor shall include with his tender full details of the method he proposes to use which shall always include

(*a*) arrangement of electrodes
(*b*) which of the techniques in clauses SI 3.094 and SI 3.095 he proposes to adopt or whether he intends to use a combination of the two
(*c*) full details of electrode spacings, and frequency of test points
(*d*) proposed number of series of tests using the expanding electrode separation technique and the expected number of tests in each series
(*e*) proposed number and extent of traverses using the lateral traversing technique.

SI 3.093. *Arrangement of electrodes*
All four electrodes shall be driven into the surface of the ground in a straight line and in a symmetrical pattern. The outer pair shall generally be those connected to the source of an alternating current and the inner pair those connected to the potentiometer.

Normally the electrodes shall be equally spaced (Wenner configuration) but if the Contractor considers that the particular circumstances warrant it, the Schlumberger arrangement may be adopted, in which the inner potential electrodes are much closer together although they maintain a symmetrical pattern.

Other arrangements may, with the Engineer's agreement, be adopted if the Contractor considers that more useful results will thereby be obtained.

SI 3.094. *Expanding electrode separation technique*
When the expanding electrode separation technique is adopted, a series of tests shall be carried out in which the electrodes are moved outwards in a straight line about a fixed central point, while the basic symmetrical pattern adopted is maintained. The first test shall be carried out with the current electrodes as close together as practicable and the expansion of the system shall continue with suitable in-

155

crements being applied until the separation between the current electrodes is sufficient to yield adequate data to the depth required. Further series of tests shall then be carried out until an adequate cover of the area under consideration has been achieved.

SI 3.095. *Lateral traversing technique*

In the lateral traversing technique the whole electrode system shall be moved along a traverse while the electrode spacing and the adopted basic pattern are kept constant. The electrode separation used shall be suited to the data the traverse is required to yield and this shall normally be found by trial and error or, where conditions are simple, by approximate calculations. Further traverses shall then be made until the whole area under consideration is covered by an adequate grid.

SI 3.096. *Presentation of results*

The results of resistivity surveys shall be presented in such a form as to show the measured variation in apparent resistivity in plan or section. Where records from boreholes are available to identify the variations in apparent resistivity, these shall be plotted on the plan or section for the purpose of correlation.

SI 3.10. *REFRACTION SEISMIC METHOD*
SI 3.101. *Apparatus*

The essential components of the apparatus for the refraction seismic test shall be as follows.

(*a*) A seismograph suited to the refraction method. This shall comprise one or more amplifying units with filters, a timing unit and a recording unit having one more channel than the number of amplifiers.

(*b*) One or more geophones (seismometers). There shall be one geophone for every amplifier unit.

(*c*) Leads to connect the geophones to the seismograph.

(*d*) A means of providing an elastic pulse near the surface of the ground. This may be a small charge exploded below the surface of the ground or a hammer blow.

The Contractor shall state at the time of tender which type of apparatus he intends to use. He shall also give a full description of it together with a statement of his experience of it for the type of work required by the Particular Specification.

SI 3.102. *Method of carrying out test*

The object of the refraction seismic method is to build up a graph showing the relationship between the time taken for the effects of an elastic pulse to reach any point on the surface of the ground and the distance of that point from the source of the pulse. The geophone spacing and the total line length (seismometer spread) shall be carefully chosen by the Contractor so that the graph has an adequate number of points on the branches corresponding to all refractors of interest. When practicable the geophones shall be placed in small holes in the ground to protect them from wind. These holes shall be made such that the geophones are a tight fit in order to maximize contact. Where peat occurs at or near the surface of the ground, the Contractor shall employ marsh geophones to ensure good ground coupling.

Where explosives are used to produce the elastic pulse, they shall be placed at the bottoms of shaft holes bored by a hand auger or other approved means. The charge shall be placed sufficiently deep to keep the breaking of the surface by the explosive to a minimum. If necessary sandbags shall be placed over holes which cannot be sunk to a reasonable depth. Where the interface between two different materials slopes, each seismometer spread shall be carried out twice, once with the source of elastic impulse at one end of the spread and once at the other end. When the seismic tests are being undertaken to determine the depth of an irregular rock surface, the Contractor shall make recommendations to the Engineer as to the number, positions and depths of such boreholes he considers necessary for adequate control of the geophysical survey. When the wave velocity through the bedrock cannot be determined from normal refraction seismic methods, it shall be established directly by making seismic measurements between two outcrops or between two boreholes where the depth to rock is known.

At the time of tender the Contractor shall give full details of the way in which he intends to carry out the tests which shall include

(a) proposed geophone spacing

(b) proposed seismometer spread

(c) proposed number of spreads

(d) proposed method of directly measuring the wave velocity in bedrock, if applicable

(e) details of any boring which will be undertaken as part of the investigation.

SI 3.103. *Presentation of results*

The results of refraction seismic surveys shall be presented in such a form as to show the measured variation in seismic velocity in plan or section. The depths or levels of refracting horizons shall be shown in the form of contours on plans, or on sections. Where boreholes are available to identify the refracting horizons, these shall be plotted on plans or sections for purposes of correlation.

SI 3.11. *CONTINUOUS REFLECTION SEISMIC PROFILING*
SI 3.111. *Apparatus*
The essential components of the apparatus for the continuous reflection seismic profiling test shall be as follows.

(*a*) A seismograph similar to that specified in clause SI 3.10 but suited to the reflection technique.

(*b*) A pressure sensitive seismometer which can float freely in water (hydrophone).

(*c*) A source, normally electrical, which can float freely in water, to produce trains or pulses of sound waves. The frequency of the sound source shall be selected from considerations of the expected geological conditions.

(*d*) A suitable motor launch to contain the seismograph and to tow the hydrophone and the sound source.

(*e*) Some means of fixing the position of the launch at any time while the tests are being carried out.

The Contractor shall state at the time of tender which type of apparatus he intends to use. He shall also give a full description of it, together with a statement of his experience of it for the type of work required by the Particular Specification.

SI 3.112. *Method of carrying out test*
The hydrophone and the sources of the sound waves shall normally be towed at a suitable distance behind the launch while it makes a series of runs in two or more directions over the area being investigated. The position of the launch shall be fixed at intervals during each run in such a manner that it can be linked to the seismic records. At the time of tender the Contractor shall give full details of how he proposes to carry out the tests which shall include

(*a*) proposed number, length and alignment of runs

(*b*) proposed spacing between hydrophone and source of sound waves and general distance that these two will be behind the launch

(*c*) method of determining the velocity of sound in the materials concerned or, if an assumed value is to be used in the interpretation of the results, what it is and what justification there is for it

(*d*) method to be used to fix the position of the launch during the tests.

This information is intended to indicate the scope of the work proposed but the Contractor shall not be held rigidly to this programme should conditions on Site call for modification of plans.

SI 3.113. Presentation of results

The results of reflection seismic surveys shall be presented in the form of contour maps or profiles showing the observed variations, seismic velocity and the positions and depths of any reflecting horizons. Where boreholes are available, the records shall be plotted on plans or sections for purposes of correlation with the geophysical survey.

SECTION SI 4. LABORATORY TESTING

SI 4.01. TEST PROCEDURE

The Engineer shall decide in conjunction with the Contractor what testing shall be carried out and shall provide the Contractor with a schedule of the required tests. Laboratory tests on soils shall be undertaken in accordance with BS 1377 where applicable.

SI 4.02. REPORTING TEST RESULTS

The following information shall be submitted in the report on laboratory tests

(*a*) *moisture content determination*

moisture content expressed as a percentage of the dry weight of the soil to two significant figures

(*b*) *sieve analysis*

(i) cumulative percentage by weight of soil passing BS test sieves after wet sieving and drying, plotted on a particle size analysis diagram

(ii) weight of sample tested

(*c*) *particle size analysis by sedimentation*

(i) cumulative percentages as for sieve analysis plus percentage less than 0·2 mm, 0·006 mm and 0·002 mm plotted on a particle size analysis sheet

(ii) loss on pretreatment to nearest 1%

(iii) weight of sample tested

(*d*) *plasticity indices*

(i) history of sample, i.e. natural state, air dried or oven dried

(ii) method used to obtain results

(iii) percentage of material passing BS 420 μm sieve

(iv) liquid limit expressed to nearest 1%

(v) plastic limit expressed to nearest 1%

(vi) plasticity index

(*e*) *undrained triaxial compression test*

(i) dimensions of test specimens (mm)

(ii) bulk density (Mg/m^3) to two decimal places

(iii) natural moisture content to nearest 1%

(iv) cell pressure (kN/m^2)

(v) rate of compressive strain to nearest 1%

(vi) maximum principal stress difference (kN/m^2)

(vii) time to failure

(viii) deviator stress/strain curve

(ix) Mohr's circle diagram for set of three tests and shear strength parameters

(x) orientation of specimen with respect to vertical

(xi) sample description

160

(xii) plasticity index

(xiii) sketch to show mode of failure of specimens

(*f*) *unconfined compression test*

(i) dimensions of specimen (mm)

(ii) bulk density (Mg/m^3)

(iii) moisture content to nearest 1%

(iv) compressive strength to the nearest 2 kN/m^2 for values up to 50 kN/m^2, to the nearest 5 kN/m^2 for values of 50–100 kN/m^2 and to the nearest 10 kN/m^2 for values of over 100 kN/m^2

(v) shear strength (kN/m^2) to two decimal places

(*g*) *consolidation test*

(i) initial and final thickness of specimen to nearest 0·002 mm

(ii) initial moisture content

(iii) initial bulk density (Mg/m^3) to two decimal places

(iv) specific gravity of soil particles

(v) graph of voids ratio versus logarithm of applied effective stress

(vi) graph of compression (mm) versus square root of time or logarithm of time (min)

(vii) coefficient of compressibility (m^2/MN) for a minimum of four pressure increments including the pressure increment of 100 kN/m^2 in excess of the overburden pressure

(viii) compression ratios and coefficient of consolidation ($m^2/year$) for a minimum of four pressure increments including the pressure increment of 100 kN/m^2 in excess of the overburden pressure

(ix) Orientation of specimen with respect to vertical

(x) detailed description of specimen

(*h*) *specific gravity of soil particles*

specific gravity of soil particles to the nearest 0·01

(*i*) *sulphate content of soil*

water soluble sulphate content of the soil expressed as percentage sulphur trioxide or in grams per litre when determined from a 2:1 aqueous extract

(*j*) *sulphate content of groundwater*

sulphate content of groundwater expressed as parts per 100 000

(*k*) pH value accurate to 0·1

If bulk density and maximum and minimum density tests are required they shall be reported in Mg/m^3 to two decimal places.